common denominators

Cultivating Engagement and Belonging in Secondary Mathematics

Jennifer A. Lenhardt

Solution Tree | Press
a division of Solution Tree

555 North Morton Street
Bloomington, IN 47404
800.733.6786 (toll free) / 812.336.7700
FAX: 812.336.7790

email: info@SolutionTree.com
SolutionTree.com

Visit **go.SolutionTree.com/mathematics** to download the free reproducibles in this book.

Printed in the United States of America

FSC
www.fsc.org
MIX
Paper | Supporting responsible forestry
FSC® C008955

Library of Congress Cataloging-in-Publication Data

Names: Lenhardt, Jennifer A., author. | Seeley, Cathy L., editor. | Bay-Williams, Jennifer M., editor.
Title: Common denominators : cultivating engagement and belonging in secondary mathematics / Jennifer A. Lenhardt ; Cathy Seeley (series editor) ; Jennifer Bay-Williams (series editor).
Description: Bloomington, IN : Solution Tree Press, [2024] | Includes bibliographical references and index.
Identifiers: LCCN 2023045836 (print) | LCCN 2023045837 (ebook) | ISBN 9781960574343 (paperback) | ISBN 9781960574350 (ebook)
Subjects: LCSH: Mathematics--Study and teaching (Secondary) | Culturally relevant pedagogy.
Classification: LCC QA16 .L46 2024 (print) | LCC QA16 (ebook) | DDC 510.71/2--dc23/eng/20240205
LC record available at https://lccn.loc.gov/2023045836
LC ebook record available at https://lccn.loc.gov/2023045837

Solution Tree
Jeffrey C. Jones, CEO
Edmund M. Ackerman, President

Solution Tree Press
President and Publisher: Douglas M. Rife
Associate Publishers: Todd Brakke and Kendra Slayton
Editorial Director: Laurel Hecker
Art Director: Rian Anderson
Copy Chief: Jessi Finn
Production Editor: Madonna Evans
Copy Editor: Evie Madsen
Proofreader: Sarah Ludwig
Text and Cover Designer: Julie Csizmadia
Acquisitions Editors: Carol Collins and Hilary Goff
Assistant Acquisitions Editor: Elijah Oates
Content Development Specialist: Amy Rubenstein
Associate Editor: Sarah Ludwig
Editorial Assistant: Anne Marie Watkins

GROWING THE MATHEMATICIAN IN EVERY STUDENT COLLECTION

Consulting Editors: Cathy L. Seeley and Jennifer M. Bay-Williams

No student should feel they're "just not good at math" or "can't do math"!

Growing the Mathematician in Every Student is a collection of books that brings a joyful positivity to a wide range of topics in mathematics learning and teaching. Written by leading educators who believe that every student can become a mathematical thinker and doer, the collection showcases effective teaching practices that have been shown to promote students' growth across a blend of proficiencies, including conceptual development, computational fluency, problem-solving skills, and mathematical thinking. These engaging books offer preK–12 teachers and those who support them inspiration as well as accessible, on-the-ground strategies that bridge theory and research to the classroom.

Consulting Editors

Cathy L. Seeley, PhD, has been a teacher, a district mathematics coordinator, and a state mathematics director for Texas public schools, with a lifelong commitment to helping every student become a mathematical thinker and problem solver. From 1999 to 2001, she taught in Burkina Faso as a Peace Corps volunteer. Upon her return to the United States, she served as president of the National Council of Teachers of Mathematics (NCTM) from 2004 to 2006 before going back to her position as senior fellow for the Dana Center at The University of Texas. Her books include *Faster Isn't Smarter* and its partner volume, *Smarter Than We Think*, as well as two short books copublished by ASCD, NCTM, and NCSM: (1) *Making Sense of Math* and (2) *Building a Math-Positive Culture*. Cathy is a consulting author for McGraw Hill's *Reveal Math* secondary textbook series.

Jennifer M. Bay-Williams, PhD, a professor at the University of Louisville since 2006, teaches courses related to mathematics instruction and frequently works in elementary schools to support mathematics teaching. Prior to arriving at the University of Louisville, she taught in Kansas, Missouri, and Peru. A prolific author, popular speaker, and internationally respected mathematics educator, Jenny has focused her work on ways to ensure every student understands mathematics and develops a positive mathematics identity. Her books on fluency and on mathematics coaching are bestsellers, as is her textbook *Elementary and Middle School Mathematics: Teaching Developmentally*. Highlights of her service contributions over the past twenty years include serving as president of the Association of Mathematics Teacher Education, serving on the board of directors for the National Council of Teachers of Mathematics and TODOS: Mathematics for ALL, and serving on the education advisory board for Mathkind Global.

For Stella and Nico,

May your experiences in and out of school
forever reinforce that you belong.

I love you so big.

Acknowledgments

Where we are today and the book you are about to read are both the result of countless interwoven physical, emotional, even spiritual happenings. It feels nearly impossible to capture the totality of this interconnectedness, and I know these acknowledgments are incomplete.

Beginning with the physical space I occupy, I acknowledge the Indigenous peoples whose land was stolen and who were removed by threat or by force from the Willamette Valley in Oregon. The 1855 negotiations between government representative Joel Palmer and the Tualatin Kalapuya resulted in what is known as the *Willamette Valley Treaty* and led to the majority of Kalapuyans, Santiam, Tualatin, Yamhill, Ahanchuyuk, Lackmiute, Mary's River (Chelamela), Mohawk (Pee-you), Winfelly, and Calapooia deciding to confederate, and the establishment of reservations in the Willamette Valley (Lewis, n.d.b). In 2023, members of Willamette Valley Kalapuyans, Molala, and Clackamas Chinook people remain members of the Confederated Tribes of Grand Ronde community. Members of the Confederated Tribes of Grand Ronde are integral to our community, actively collaborating with local and state governments; promoting and sharing their tribal knowledge, culture, and language; and providing generous monetary support for our community (Lewis, n.d.a). The land on which I reside cannot be separated from my existence or experiences. Acknowledging this is one small way to recognize my history in a way that affects my current and future decisions.

This leads to the heart-foundations of this book. I want to acknowledge and thank my parents, Alison Kelley and Rick Lenhardt, for the life and love you built and nurturing you continue to offer me. From the late-night conversations discussing the challenges and opportunities in our communities to caring for my tiny humans on the warm summer afternoons while I worked on this manuscript, your support and love helped make my dream of writing a book a reality. I am forever grateful for the ways you move through this world, the example you set, and the love you share.

Thank you to my sisters, Amy, Jessie, and Catherine. You have your own ways of being in the world, and I learn from, honor, and value each of you. Our sisterhood is a place of life-giving belonging for me; a place where I am held, loved, nurtured, challenged, and encouraged. Thank you for being exactly who you are and loving me exactly as you do.

I cannot capture the name of every student, teacher, neighbor, family member, colleague, and friend who matters to my journey, my learning, and this book. Through our interactions, I have learned so much that challenges me to think, feel, and expand my perspective and understanding. Journeying alongside each of you contributed to who I am today and who I am still becoming. Thank you.

"There's a book in there, Jennifer," Jamie Cross repeatedly encouraged me after late dinners and long meetings focused on surfacing our best strategies and approaches to supporting mathematics teachers. Thank you, Jamie. You knew and believed in this book before I did. We are all beneficiaries of your guidance, advice, and encouragement.

Thank you to Holly Terpening, Dan Holden, and Kelly Harper. Your dedication to students, teaching, and storytelling is now laced throughout the pages you helped edit. Thank you to Meaghan Pavlovich for our countless discussions about mathematics teaching and learning, our wholistic experiences in this world, and your unique skill set for understanding and making connections between research and practice. Thank you for poring over every detail of this manuscript and helping me tell these stories with clarity and connected purpose.

Thank you to Cathy Seeley, Jennifer Bay-Williams, Carol Collins, Jamie Vincent, Amy Rubenstein, Madonna Evans, Rian Anderson, and the team at Solution Tree for believing in this book and supporting the journey to bring it to print.

Solution Tree Press would like to thank the following reviewers:

Taylor Bronowicz
Mathematics Teacher
Sparkman Middle School
Toney, Alabama

Jordan Edgerly
Mathematics Teacher (NBCT)
Waukee Northwest High School
Waukee, Iowa

Amber Gareri
Instructional Specialist,
Innovation and Development
Pasadena ISD
Pasadena, Texas

Kelly Hilliard
Mathematics Teacher
McQueen High School
Reno, Nevada

Sheryl Walters
Instructional Design Lead
Calgary, Alberta, Canada

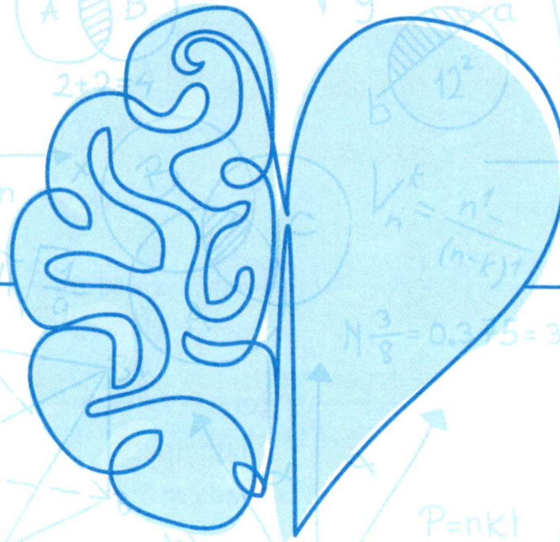

Visit **go.SolutionTree.com/mathematics**
to download the free reproducibles in this book.

Table of Contents

Reproducibles are in italics.

About the Author

Jennifer A. Lenhardt (she/her), MEd, is a senior manager for professional services with Houghton Mifflin Harcourt, where she integrates her experience as a teacher and her research interest into equity and culturally responsive teaching with professional learning. Formerly, Jennifer taught mathematics intervention at a middle school in Oregon, working collaboratively with educators across the district to launch and refine approaches to Tier 3 interventions and supporting multilingual mathematics learners. Jennifer's education and passion focus on middle and high school students who have yet to see learning mathematics as a source of joy, inspiration, or belonging. Through ongoing analysis of factors affecting the development of students' and teachers' multiple and intersecting identities, Jennifer seeks to describe how we can better understand our lived experiences in and out of school, and create the spaces necessary to flourish.

As a member of the National Council of Teachers of Mathematics (NCTM), Jennifer presented "Breaking Cycles of Failure: Reframing to Reengage Students in Learning Math" at the NCTM National and California Mathematics Council conferences. During her presentation at the Model Schools Conference ("Amplifying Opportunity: Turning Up the Volume on Agency, Equity, and Courageous Teaching"), she shared her passion for reconsidering how we frame mathematics learning challenges and next steps teachers can take to operationalize their learning. She writes for Houghton Mifflin Harcourt's *Shaped* blog (https://hmhco.com/blog; posts include "Math Talk: Making the Most of Math Class Discussions" and "Who's Doing the Work? Developing Math Muscles") and is a co-creator and former cohost of the *Method to the Mathness* podcast.

Jennifer earned a bachelor's degree in political science with departmental honors from the Robert D. Clark Honors College at the University of Oregon, and a master's degree in education with endorsements in secondary mathematics and English for speakers of other languages (ESOL) from the University of Oregon.

To book Jennifer A. Lenhardt for professional development, contact pd@SolutionTree.com.

lemf(x)=±∞
x→n
T=2π√(1/g)
2+2=4
b³=(a+b)(a¹-ab+b¹)
yo=bxn
Δ3=Δmc²
N 3/8=0.375=37.5%
P=nkt
f(-x)=a(-x)+b=-(ax-b)
a³+b³=(a+b)(a¹-ab+b¹)
E=mc²
=0.375=37.5%
(-1)ⁿ x²ⁿ / (2n)
FUNCTION
x/a²+y/b²-z/c²=1
P=F/S

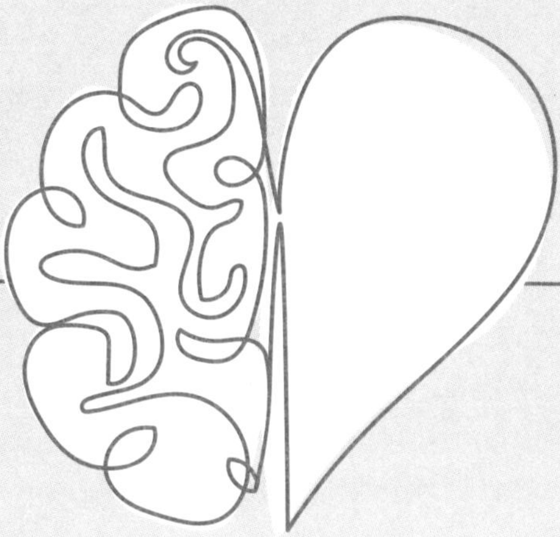

A possibility was born
the day you were born,
and it will live as long
as you live.

—Marcus Solero

INTRODUCTION

Teacher, Where You From?

The early weeks of September are the finest the Oregon summer has to offer. The scent of warm, ripe blackberries drifts on the breeze, rustling the leaves that will soon change. Back-to-school season in the Willamette Valley is *magical*, especially for those like me who cherish the beginning of the school year.

The familiar preparations—arranging desks, organizing materials, studying student profiles, and writing initial lesson plans—lead to that first bell and set the stage for what promises to be a magical year, filled with student learning and inspirational mathematics lessons! (A girl can dream, right?) As I stand at the classroom door welcoming middle school mathematics-intervention students to Portable 137, I project joy and excitement before a million bits of reality confront me.

Students file in and take their seats. The bell rings, and I break the nervous silence with an overzealous, "Well hello, everyone! I'm Ms. Lenhardt. Welcome!" To start creating a sense of belonging right away, I engage students in some getting-to-know-you discussions. I begin by asking students, "What was a highlight from your summer?" and "What are you looking forward to learning this year?" Diverse tones from a variety of languages and accents intermingle to create an intricate, delicate melody. A student asks me, "Teacher, where you from?" This is a question (or some variation of it) I've heard from students before. The face and attitude that usually accompanies this question is of the leaned-back, arms-crossed, chin-tilted variety. What they are really asking, of course, is, "Who the heck are you?" Y'all—it is just so precious how young adults express their interest in getting to know others! Don't you just love middle school?

Allow me to connect this question to the broader set of typical student sentiments I hear, many of which I'm sure you have heard as well. Each school year, as we launch into the academic work before us, I've noticed a pattern of student statements that pique my curiosity. Figure I.1 shows some of the student declarations that often reverberate off the walls of our beloved classrooms and that speak to the need to foster engagement and belonging.

Figure I.1: Things students say in mathematics class.

It's All About the Denominator

Fractions can help us visualize and understand the work of cultivating engagement and belonging in the mathematics classroom. I'll use the *part-whole definition of fractions* for this, which states that in a fraction, the *numerator* represents a part, while the *denominator* represents the whole.

Throughout this book, I will share stories of both students and teachers, and the things we say about ourselves and our relationship with mathematics. These stories are the numerators; they are the part we hear on the surface. The denominator represents the whole, the entirety of the human story and experience. I'll then introduce practices, structures, tools, and strategies for gaining insight into the whole (the denominator). Visualize the fraction bar as the divider of an iceberg, separating what's visible above from the mass underwater. Figure I.2 illustrates that many numerators are parts of the whole-student and teacher experience, which has two consistent factors: (1) engagement and (2) belonging.

$$\frac{\text{Part}}{\text{Whole}} = \frac{\text{What our students say to us, what we say to each other}}{\text{(Engagement) (Belonging)}}$$

Figure I.2: What students say is only the tip of the iceberg.

Bear with me for a moment. Let's take this a layer deeper by using *factor trees* to visualize key components of engagement and belonging. Figure I.3 illustrates how we use factor trees (also called *number mountains*) in a mathematical context to visualize how to break apart composite numbers into smaller factors. This serves us in a variety of mathematical application contexts.

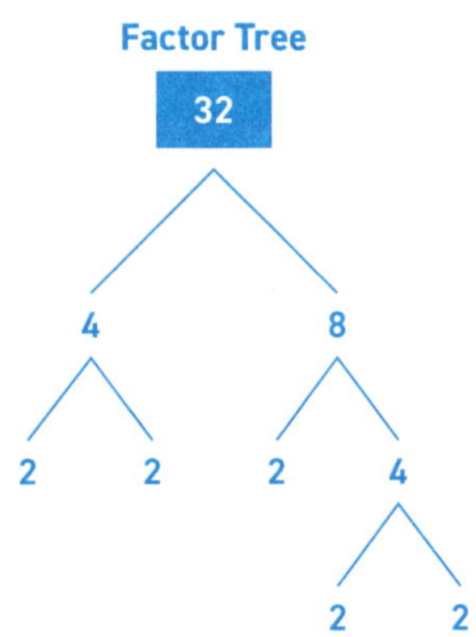

Figure I.3: Visualizing factors of composite numbers.

Let's translate this structure to help visualize factors of engagement and belonging. Figure I.4 (page 4) illustrates some of the factors that contribute to engagement and belonging. Notice how the factors of engagement and belonging are less linear than, for example, the factors of 32 (see figure I.3), and the intentional listing of these factors under both engagement and belonging.

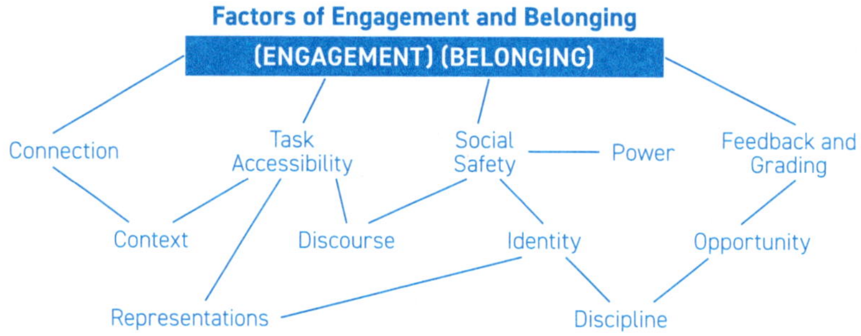

Figure I.4: The many factors of engagement and belonging.

While we may think of certain factors, such as *task accessibility*, as simply a factor of engagement, I posit it is a factor of both engagement and belonging. Through creating and facilitating accessible tasks, we foster belonging *and* motivate engagement, not one or the other. As we factor these composite ideas and develop skills to cultivate each, we become equipped for a variety of instructional and professional applications.

Keep in mind, there may be multiple factors driving the parts we see above the fraction bar. When we address any one of those factors, we are effectively changing the whole.

The aim of this work is to build our conceptual understanding through investigation of engagement and belonging factors so we can intentionally take steps to cultivate them with our secondary students and across our profession. This work is necessary for student learning in mathematics specifically and, more broadly, for sustainable teacher practice.

The Case for Engagement and Belonging

Something strikes me the moment I hear a student ask, "Teacher, where you from?" I pause because *I just don't think that's what they're really asking about.* My students seek information and details about others so they can piece together the puzzle of who a person is in relation to who *they* are. Our students want to know, from the moment they enter class, if they can establish our trustworthiness by discerning where we are from. To my ears, this surface-level question sounds an awful lot like, "Teacher, is there room for me here?" This student question about my origin may seem like a typical exchange for a getting-to-know-you activity. However, there are crucial, research-informed reasons to believe there may be more going on here.

Professor of education Rochelle Gutiérrez's (2009, 2013, 2018) work to advance the discourse and practice of rehumanizing mathematics articulates how we are responsible for creating spaces where our identities and the identities of our students all belong. By this I mean we all can bring our complex, multiple, and intersecting identities—indeed our whole selves—to mathematics (Gutiérrez, 2009, 2013, 2018). We must be intentional with the role we play in nurturing our own identities and those of our students, and ensure we don't accidentally trample over the identities developing in our classrooms.

Additionally, research points to the relationship between success in algebra and a sense of belonging: "not only is a sense of belonging to mathematics a significant predictor of middle school students' algebra learning but it was the only significant predictor of the motivation and belief measures taken" (Barbieri & Miller-Cotto, 2021, p. 7). Just as number sense inextricably links to proficiency generally, so too does belonging inextricably link to academic success and well-being (Hammond, 2015).

Your lived experiences are research too. You have likely experienced the overwhelming fatigue that comes with attending to each symptom (or above-the-surface need) that presents itself. Countless student requests for restroom breaks, interruptions to instruction, side-chatting with peers, and so on take the wind right out of your sails. Because teachers are charged with leading a group of students forward and are often accountable for their academic success and well-being, we can easily start to feel frustrated with—or even powerless over—the circumstances that delay or even fully stop progress. We need a more sustainable model!

Conventional approaches to instruction must shift as a result of our deepening understanding of how engagement and belonging are intertwined and necessary. Rather than addressing each above-the-surface expressed need, we must develop our skills to identify the factors of what we're hearing and seeing. What is really motivating what students say or do here? We need structures for identifying patterns and trends. We need guiding questions and intentional reflective practice to build our conceptual understanding. This leads to the question, How do we go about making the instructional changes that will meet the actual needs of our students instead of treating the symptoms? This book aims to both identify and understand the factors, and provide applicable tools and strategies.

About This Book

Common Denominators, which is for middle and high school mathematics teachers, is a collection of stories braided together with research-informed strategies and tools. These are the real stories of my own learning, my students' learning, and the ongoing development of our multiple, intersecting identities. This book is part of my ambitious journey to establish and protect equitable learning in classrooms. It is a book about how you convey to students they belong in mathematics class and ultimately in *any* class. It is a book that makes space for *your* stories, reflections, ponderings, and thoughts about teaching and learning.

You don't have to search far to see how stories and storytelling are integral to our survival. Stories make us *feel*. They resonate with our experiences and echo in the chambers of our souls. They are the connective fibers between individual and collective journeys, and they remind us *we are humans* living alongside other humans. The stories we tell reflect our identities, and, as we continually edit and rewrite those stories, we are, in turn, empowered to craft stories reflective of who we are becoming. This is true for you, for me, and for each student. I hope this book inspires you to write, edit, revisit, and revise the story you want to write with each classroom moment.

This book is not a formula sheet. There are no magic answers or quick sets of predefined steps that will cure our students or schools of all that ails them. You'll notice throughout the chapters, there isn't a T-chart that lists *matters of engagement* on the left and *matters of belonging* on the right because of the interconnectedness of these strategies and tools. While we often crave the certainty a formula provides, the work of teaching—indeed of being human—is less about having the formula and more about understanding the variables and how they operate together.

The aim of this book, rather, is to provide the scaffolding we need to build conceptual understanding and activate strategies to cultivate engagement and belonging as we teach mathematics. There is nothing *simple* about classroom management or instructional design. And at the same time, it isn't magic either. There are structures to examine, routines to study, and skills to develop as we grow and learn alongside our students. Tools to help students navigate intense emotions meet a belonging need and serve as a mechanism for engagement. Our collective synergy propels us forward as we learn to see our students as fully human and refuse to passively (or actively) participate in leaving students emotionally or academically stranded.

What's in This Book

Each chapter launches with a story from my time teaching mathematics intervention or consulting and coaching teachers from Hawaii to Maine. We will then consider the question, What's really going on here? by diving beneath the surface of the story (part) and investigating what might be at its root (whole). I will share tools I have learned or created that cultivate student engagement in learning mathematics and their sense of belonging as learners and doers of mathematics. Please note, names and identities in this book have been changed to protect privacy.

Additionally, because we bring our whole selves to this work, I also share strategies and tools for developing yourself. You will find Pause and Ponder prompts throughout the chapters, which will help guide your own analysis and application of the strategies and tools. At the end of each chapter, you'll find a section called *To Sum It Up*, which will connect back to the initial story and tie together the threads woven throughout. Each chapter also includes a reproducible application guide to help you connect the chapter's themes to your practice. The application guides are two-dimensional tools for three-dimensional work. Consider bolstering your experience with this content by creating a concept map or another visualization of the interconnectedness of the chapter themes. The book also includes two appendices with helpful resources for diving further into the work of cultivating engagement and belonging.

Common Denominators is divided into two parts. Part 1 contains five chapters that explore some of the most common things our students say to us. Part 2, which includes three chapters, is dedicated to the things we teachers often say to one another.

- Chapter 1 focuses on finding a name to accurately describe behaviors, noticing the patterns in student interactions among one another and us, and crafting an approach that disrupts cycles of failure.
- Chapter 2 shares strategies for discerning the source of students' frustration and provides guidance for answering the questions students are not yet asking.
- Chapter 3 focuses on power dynamics and explores options to reframe zero-sum interactions and restore a learning environment designed to help students flourish.

- Chapter 4 is about the de-escalating power of empathy and provides recommendations for how to model and teach emotional literacy.
- Chapter 5 provides guidance for helping students and teachers differentiate between who they are and what they can do, and avoid burnout that can lead to frustration with mathematics.
- Chapter 6 describes alternatives to traditional approaches to grading and the long-term implications of *F*s.
- Chapter 7 provides a framework for examining and possibly deconstructing engrained beliefs that may not be serving us well.
- Chapter 8 details how to support yourself and other teachers as you encounter the challenges in our profession.

My hope is that, through reading and interacting with the stories in this book, you can better understand what may be beneath the surface of cyclical learning struggles. I hope, in some way, to be your curiosity mentor as we explore how to disrupt patterns that keep us stuck, and learn how to help students reclaim the learning opportunities that rightfully belong to them. Perhaps more importantly, I hope this book will help you unearth *why* it is so crucial to fully understand our profound and awesome impact on each student.

Struggle and triumph punctuate the teaching journey, and also the journeys of students. We will find one another at the bustling intersection of engagement and belonging, and write our own narratives as we empower others to write theirs. My sincerest hope is that the stories, strategies, and resources in this book resonate with you and support your work in ways that reassure: you are not alone on this wild journey of impacting young lives. You matter. Your incredible, unbounded dedication to this craft matters. This dedication is a part of your story and, indeed, your very identity. We need you in this space, and I am so very glad you are here.

0.375 = 37.5 %
T=2π√(1/g)
E=mc²
Δ3=Δmc²
a³+b³=(a+b)(a²-ab+b²)
∫ 1/x dx = ln|x|+c
P=nkt
P = F/S
2+2=4
y0=bxn
f(-x)=a(-x)+b=-(ax-b)
x/a² + y/b² - z/c² = 1
FUNCTION

PART 1

What Our Students Say to Us

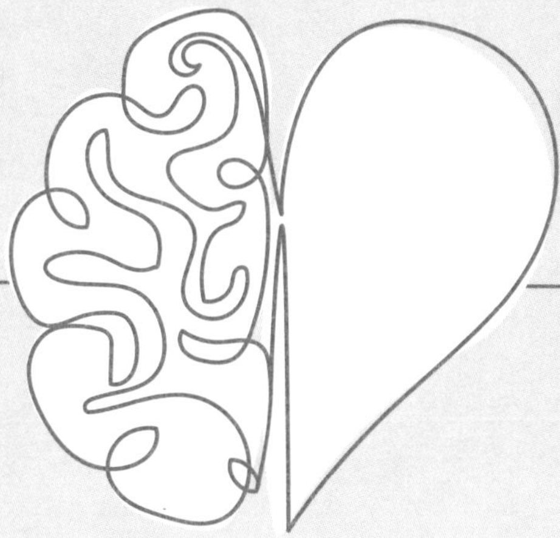

In relationships,

children develop models

about how they themselves

fit into the world around them,

and how relationships work.

They learn whether they can trust others

to see and respond to their needs,

and whether they feel connected and protected

enough to step out and take risks.

In short, they learn whether relationships

will leave them feeling alone and unseen;

anxious and confused;

or felt, understood, and securely cared for.

—Daniel J. Siegel
and Tina Payne Bryson

CHAPTER 1

I Don't Have a Pencil

"I don't have a pencil," Jaidyn says about twelve minutes into seventh-period mathematics class.

"What have you been doing this whole time?" the teacher replies, somewhat incredulously.

The other students chuckle, snicker, and snort without looking up. This is the norm in this class. Jaidyn waits until just the right time to capture the spotlight as she pronounces to her peers, in one way or another, that she is decidedly off task. It gets a good laugh, every time.

"Jaidyn, when will you start taking some responsibility? I gave you a pencil yesterday!"

"Oh, you mean this one?" Jaidyn says as she pulls the pencil from her backpack.

"Oh, no you didn't!" her classmates egg her on. This show never gets old! Mathematics class with Jaidyn in the house is entertainment for sure.

PAUSE AND PONDER

What do you suppose the teacher said or did next? What would you say or do next?

Would your responses be the same, or would they change depending on the circumstances or your knowledge of Jaidyn? How might facets of her identity (gender expression, race, ethnicity) or past (previous school discipline events) influence your reaction?

The next part is like a *Choose Your Own Adventure* (www.cyoa.com) ending. Let's imagine a couple of potential endings to this story.

- **Ending one:** You say, "This is not the first time you've disrupted our learning. Consider this your warning. Get to work." Jaidyn mumbles an insult under her breath, then looks down, and stares blankly at her paper.
- **Ending two:** You say, "Y'all won't think this is very funny when you're taking your test tomorrow. Those of you working right now have a better shot at passing that test than those of you messing around."
- **Ending three:** You roll your eyes and laugh. "You got me, lady. So clever as usual," you say with a smile. As you walk over to Jaidyn's desk, you ask, "Tell me how you think you'd get started on this."

These are not the only ways this scene might end, but they're examples of what I've seen and heard in secondary mathematics classes. Who have you been in this story? Have you ever been Jaidyn? Have you been the teacher? Jaidyn's classmate? I've been Jaidyn, fiercely afraid of failing and leveraging any tactic in my arsenal to avoid encountering that failure, *especially* if the failure might be in front of my peers. Humor is my go-to when I'm uncomfortable or when things get tense. Although on the surface it can seem like an attention-grabbing tool, it is my most effective weapon of *mass distraction*. People start laughing, and the story I tell myself is if they're laughing, they won't notice I'm confused.

I have been the teacher too. And I will bravely (not proudly) admit that over the course of my teaching, I have offered up each one of those endings. My response depended on what mood I was in, how many times I had been sassed that day, or if my formal observation was happening right at that moment. Many factors impacted what I *heard* and how I *responded*—and as you can imagine, each response had a markedly different impact on the classroom culture, level of student engagement, and ultimately, learning of mathematics.

What's Really Going on Here

You may recall the game *tug-of-war* from your childhood; it's a game where half the members of a group hold one end of a long rope, and the other half holds the other end. The two sides face off against each other and pull with all their might to try to overpower the other team. It often results in rope burns and tension between the teams. Despite the injuries and ill will, I still enjoyed playing the game. This natural tendency to compete is something to consider when deciding how to interact with students.

POWER STRUGGLES

Many student-teacher interactions can feel like a game of tug-of-war. It doesn't take long to realize how *wildly ineffective* it is for a teacher to play tug-of-war with students. Engaging in rounds of tug-of-war is contrary to co-creating an environment that fosters engaging, lasting learning. It breaks down any sense of belonging and disintegrates into a power struggle. For example, consider Jaidyn and the pencil situation (see page 13). Figure 1.1 shows how the different scenarios correlate with the tug-of-war framework.

The Scenario	The Tug-of-War Framework
"I don't have a pencil," Jaidyn says.	"Let's play tug-of-war."
The class chuckles, snickers, and snorts under their breaths.	Audience secured.
"Jaidyn, when will you start taking some responsibility? I gave you a pencil yesterday!"	"You bet; I love tug-of-war."
"Oh, you mean this one?" Jaidyn says as she pulls the pencil from her backpack.	"I'm gonna win."
"Oh, no you didn't!" her classmates egg her on.	Class versus teacher

Figure 1.1: Jaidyn's pencil scenario and the tug-of-war framework.

I am reminded of something one of my professors said: "Never, ever engage in a power struggle with a student. You'll lose every time."

Power struggles and student-teacher confrontations do not always look or sound like a traditional showdown. Sometimes these interactions masquerade as innocent questions, eye rolls, or heads down on desks. "Let's play tug-of-war" is an invitation that might sound like, "I don't have a pencil," or "When does this class get out?" It might even sound like, "I hate math," or "When am I ever going to use this?"

Note that a small percentage of the time, this is *actually about* simply needing a pencil. (We will come back to that later in the chapter, because accessing the materials students need can be a real struggle, and I hope to offer a useful idea or two.) For Jaidyn, this wasn't about a pencil, and I would suggest for many of your students, it isn't either. Although some might identify these power struggles as anything that pushes your buttons—and I see that perspective—I would add these *power struggles* are any action, statement, verbal or nonverbal cue, perceived attitude, or micromoment that poses the core questions, *Do you see me? Do I matter?* Look at figure 1.2 to revisit our three possible endings and consider how each one answers these central questions.

The Scenario	The Tug-of-War Framework
Ending one: "This is not the first time you have disrupted our learning. Consider this your warning. Get to work."	Jaidyn pulled. The teacher pulled harder. It is unclear who the winner is because this exchange leaves everyone feeling at a loss. The message from the teacher: **I see you, but for the record, I'd rather not. And you don't matter.**
Ending two: "Y'all won't think this is very funny when you're taking your test tomorrow. Those of you working right now have a better shot at passing that test than those of you messing around."	In an effort to equalize the tug-of-war, the teacher has pulled some students to her side. Still, two teams are pitted against each other. The message from the teacher: **I see you. And your mattering is contingent on your performance.**
Ending three: Jaidyn's teacher rolls her eyes and laughs. "You got me, lady. So clever as usual," she says with a smile. As she walks over to Jaidyn's desk, she asks, "Tell me how you think you'd get started on this."	The teacher sets down her side of the rope and walks over to Jaidyn's side. Standing behind her student, grabbing hold of the same side of the rope, the teacher says, "I see what you see. I'm with you. Let's pull." The message from the teacher: **I see you. And you matter.**

Figure 1.2: Three endings to Jaidyn's pencil scenario.

Choosing to set down your side of the rope doesn't mean you *can't* win. You don't choose to put down the rope because you are not strong enough, a poor litigator, or a less-capable opponent. It is an intentional move you make to protect the learning environment and the fragile relationships under constant stress in schools. It is a move you make because you are committed to students knowing they are seen and they matter, and it is a critical aspect of cultivating belonging.

Thus begins the hard work of figuring out *how* to put down your rope and see things from a student's perspective when you are feeling run down, overwhelmed, annoyed, angry, or distracted. It is hardest to prevent defaulting to a tug-of-war reflex when students push your buttons. When I'm in my grown-up teacher mind, putting down the rope and not playing tug-of-war is easy peasy. It is when I am *anything* less that I can find myself defaulting to the tug-of-war dynamics so destructive to the other affirming, identity-building work I'm trying to accomplish. It takes a lot of practice and many setbacks to develop the skills to *not* engage in tug-of-war.

PAUSE AND PONDER

Who are the students *literally* asking for pencils in your class? Why do you suppose that is?

Who needs a pencil but isn't asking? Why might that be?

Who are the students *figuratively* asking for a pencil—the students inviting you to play tug-of-war in some form or another?

FEAR OF FAILURE

It is necessary to examine failure, or a sense of it, as a part of students' learning experiences as we seek to cultivate belonging in our classes. Students express a feeling of failure or impending failure differently. Some students manage the feeling of failure by actively defying learning, stomping out of class, or refusing to come to class in the first place. Others express their experience more passively, resigning themselves to the inevitable defeat they have experienced many times before. This may look like a head down on a desk, not putting forth much effort, or otherwise trying to "fly under the radar."

Cultivating belonging requires developing the skills to notice and not dismiss the behaviors that signal students' sense of failure, and determining how to intervene to disrupt those patterns and shift the narrative unfolding in our classrooms (Barbieri & Miller-Cotto, 2021). "Failure is a feeling long before it is an actual result," former First Lady Michelle

Obama (2018, p. 43) writes in her book *Becoming*. Let me say it again for the people in the back: *Failure is a feeling long before it is an actual result.*

Remember Jaidyn? Her choice of action posed the questions she asked—in not so many words—"Do you see me? Do I matter?" She was earnestly asking, "Do I belong here?" Two of the potential endings to the pencil scenario confirm for Jaidyn that she was on her own to manage her feelings and experience. In those two instances, the teacher breathed lighter fluid on the embers of a feeling she initially had the opportunity to snuff out. Compare this to what we see in the third ending. It's a teacher move that says, "In no uncertain terms, you are not a failure, and you are not alone. You *belong* here."

Let's take a look at two more pencil-related scenarios in figure 1.3 and consider the emotional undertones of the student experience.

Scenario	The Emotional Undertones
Teacher: "Where's your pencil?" **Alessandro:** "I forgot it." **Teacher:** "Does anyone have a pencil for Alessandro?" **Jorge:** "Here, man, you can have this one." **Alessandro:** "Nah, it's fine." **Teacher:** "Take it. Let's not waste any more time. Thank you for being prepared, Jorge."	The teacher, who called out to the other students that he was unprepared for class, shamed Alessandro. A student may have several reasons for rejecting the pencil a peer offers. In Alessandro's case, he doesn't want his peers to know his family struggles to afford school supplies. When the teacher insists he take the pencil Jorge offered, Alessandro dejectedly accepts it. For the rest of class, Alessandro has his head hung low or resting on his desk. He won't engage in conversations with peers or make eye contact. He's embarrassed and weary of feeling like a failure.
Teacher: "Put away your pen. Remember you have to use a pencil in mathematics class." **Harper:** "I don't have a pencil today." **Teacher:** "How are you going to do your work without one?" **Harper:** [Pause] "Uh, with my pen." **Teacher:** "But you can't use a pen in mathematics." Harper stands up, shouts an expletive, and storms out of class.	The student's efforts to engage in learning are met with teacher expectations for compliance. The student has failed to comply with the rules, and expresses her feeling of failing by storming out. Students in this position are effectively saying, "I cannot win. The cards are stacked against me, and I'll always fail."

Figure 1.3: Two more pencil scenarios.

Without intentionally working to level the academic playing field, we can often inadvertently tilt it so steeply students opt out of learning. This is what happens in both scenarios in figure 1.3. Whether passively defeated or actively defiant, these two students are not engaged in mathematical reasoning and sensemaking during the lesson. They are not forming predictions, testing ideas, or tackling challenging mathematics tasks. Because why bother, when learning starts with a hassle for the essential need for belonging and basic materials? We have a choice in these moments to take actions that convey to students how much they matter and bring them into the fold of the learning experience, rather than actions that marginalize and push them away.

Finding Solutions

Whether your pencil-related challenges are literal or symbolic, it can be helpful to learn ways to approach solving these issues. As these challenges surface, I recommend working in teams or otherwise seeking out colleagues who may have valuable insights, such as your building mathematics coach or another teacher. Such guidance and thought partnerships can come from even the least-expected places, such as dinner-table chats with friends. Allow me to share some of the lessons I learned about cultivating both engagement and belonging, plus solutions for when a challenge is literally about pencils versus when the pencils are symbolic.

WHEN THE ISSUE IS LITERALLY ABOUT PENCILS

The following are some ideas for solutions when the challenge is actually about the lack of pencils or other needed supplies.

- Ask your local office supply store manager for help. For example, stop by and say, "Hi! I teach at __________ school. We are always short on pencils in my class. Would you be interested in donating any?" I don't recommend formal donation requests. Look in the face of a manager and ask personally. If you're successful, be sure to send a thank-you note, and share with students that there are folks in the community who care about them and want to help. In doing so, you model for students, *You are worth asking for what you need, and we collectively belong in our community.*
- Let students know throughout the year that the whole class will rely on one another for materials. If students would like to donate supplies to the class, they can connect with you privately. Because, *We are not here to make a show of who donates and who does not.*

- Ask other teachers if they have surplus supplies. Some teachers have supply budgets they don't use. It's worth asking around.

WHEN THE PENCILS ARE SYMBOLIC

The following are some ideas for solutions when the challenge seems to be about something other than a pencil or other needed supplies. Notice the ways these ideas set students up for engaging in learning and cultivate a sense of belonging.

- Be preemptive by standing at the door, holding a bunch of pencils, and greeting students by name as they walk in the classroom. Say to each student, "You're going to need something to write with today. Would you like a pencil?" Because, *If you need a pencil, then this is a place where you can get what you need.*
- Get comfortable apologizing when you've marginalized students in any way; you need to make it right. It can happen in the moment or you may realize it later. If you lose your cool with students or make any move that intentionally or unintentionally marginalizes them (recall the various responses to our students in the pencil scenarios), it is apology time. Apologies can powerfully model how to name and identify feelings, take responsibility for a mistake, and rehumanize teachers as people who also make mistakes (Darling-Hammond, 2023; Hammond, 2015; Lian, Kristiawan, Primasari, & Prasetyo, 2020). Apologies are a step toward restoring the kind of emotional safety students need for learning to occur. If the incident happened in front of their peers, so should the apology. Here's my favorite apology formula. (As mathematics teachers, formulas likely feel as safe and useful for you as they do for me.)

  I'm sorry. I should not have ________________. It was the wrong thing to do. I felt ________________, and I should have handled that differently. I want you to know you matter to me and to our class.

- Keep the goal in mind. Choosing to take a big-picture perspective can help you stay forward focused and notice when an invitation to play tug-of-war is a side road you do not want to take.

PAUSE AND PONDER

What have you tried to secure pencils or other student materials when needed?

How do you respond with the pencil is symbolic? What has worked? What does not? What might you try next?

Unhooking Our *If-Thens*

Our world, internal narratives, and work as mathematics teachers is fully loaded with *if-then* statements—hooked-together assertions that if *thing A* is true, then *thing B* will inevitably be the outcome. Here are a few examples.

- If you study hard and show up to class, then you will get an *A*.
- If you obey your parents, then you will not get in trouble.
- If you draw a transversal through two parallel lines, then the alternate exterior angles are congruent.

So many statements like these are *capital-T Truths*, or at least feel that way. So many of them are *not*, or certainly are *not always*. Studying hard and showing up to class is a good start to mastering mathematical content, but there is no guarantee of that outcome. In fact, holding fast to a statement with students is exactly the kind of thing that can make some students feel like they don't belong and are doomed to failure. ("I showed up to class and studied hard, and I still failed. I'm hopeless!")

These if-thens play a role in how we interact with the people around us, especially students. It takes time and feels like work to develop the skills to recognize and unpack these sometimes counterproductive assertions when analyzing classroom interactions or figuring out how to not pick up that darn rope.

I've heard many variations of if-then explanations related to the pencil scenario, including the following.

- "If I give her a pencil, she'll never learn responsibility."
- "If they don't have a pencil, they can't do their work."
- "If I give one student a pencil, then every student will ask for one."
- "If I give out pencils, then I have to spend my classroom budget (or my own money) replenishing the stock."

Writing out our if-thens is critical. Without these words, we can't get to the bottom of our narratives and decide if it is capital-T Truth or not. This reflective work gives us a starting place to bring to the surface our own internal constructs of what we believe is true, examine it, and decide if we want to continue operating from that truth or not. This often includes engaging in a little perspective taking.

PAUSE AND PONDER

What about your *if-thens*? Are any of these examples familiar to you?

Taking a New Perspective

I would like to share how perspective taking looks for me, as it may shed light on your own if-thens. Doing so will help you as you endeavor to create a same-side dynamic in your own tug-of-war situations, where you and your students are collectively pulling against the adversity, challenge, and confusion, rather than pulling against each other.

When I find myself clinging to an if-then, refusing to let it go, I can almost always dig deeper and find *another* false if-then about which I regularly have to gut check. It is this: *If I take perspective about __________, then I am excusing hurtful or otherwise unacceptable behavior.* When I feel resistance to taking students' perspectives, it is often because I am concerned that I'm also somehow sending the message that I'm excusing a not-OK behavior. For example, when a student loses his temper (not OK), it takes intentional cognitive energy to be a perspective taker by reminding myself that taking this student's perspective does not equal excusing the outburst. What *perspective taking* does, however, is help me find a humanizing approach to my response, one aimed at restoring belonging and trust rather than one focused on punishing behaviors. Perspective taking reminds us to honor and respect others, which is essential if we are to convey to people they matter, and indeed, belong. Here are a few examples demonstrating the power of taking a new perspective to unhook an unhelpful if-then.

- **If-then:** "If I give her a pencil, she'll never learn responsibility."

  Perspective taking: There could be a thousand different reasons she needs a pencil. I want her to know she can ask for help from me, so *yes*. You bet you can have a pencil.
- **If-then:** "If they do not have a pencil, they can't do their work."

Perspective taking: How do I set up accessible mathematics learning when I've constructed a roadblock to that learning? What digital tools are available to us? What are my avenues to ensure we have materials?

- **If-then:** "If I give one student a pencil, then every student will ask for one."

  Perspective taking: I am not getting played here; I see what is happening. I hand out pencils anyway.

- **If-then:** "If I give out pencils, then I must spend my classroom budget (or my own money) replenishing the stock."

  Perspective taking: I am creative and resourceful. So are my students. We can tackle this one together. I can bring them into problem solving that directly applies to our shared learning environment.

PAUSE AND PONDER

Would your perspective taking result in any different interpretations of these examples of if-thens?

Visualize the physical orientation associated with the work about perspective taking. The only way to see another person's perspective is to be on the same side of the "rope," looking at what they are looking at, and as close as you can get to walking in their shoes. We cannot authentically say, "I see what you see," when we are squared off against our "opponents." That stance makes it impossible to be a perspective taker. The call to take another person's perspective forces the issue of "putting down your side of the rope" so you can move to a space that allows you to see what another person sees. And from there, you tackle the issue together.

To Sum It Up

Name the behaviors, notice the patterns, and craft an approach that disrupts the cycle of failure and cultivates belonging.

Your students need so much more from you than pencils. They are learning about relationships and human connections. They are learning whether they should expect others to treat them respectfully. They are learning how to mess up and make it right, what it feels like to be in the margins, and what it feels like to matter.

You probably know much of this, and chances are, you regularly execute your supportive role in developing students' mathematical identities. For the times that seem mysterious (like when you cannot seem to wriggle out of a negative interaction loop from which you and your students would like to be free), consider if the feeling of failure is potentially masquerading as something else. Examine the power struggle circumstances and tug-of-war opportunities. Engage in perspective taking to unhook your if-thens and craft approaches that achieve your aims to cultivate the common denominators of engagement and belonging. In attempting to see beneath the surface of "I don't have a pencil," choose to interact in ways that speak against the never-ending noise of *You are not enough*. For students longing to know, we get to say every time, "I see you. You matter. You belong here."

Chapter 1 Application Guide

Use the reproducible "Chapter 1 Application Guide" to gather your thoughts and make connections between the themes and tools in chapter 1 and your teaching practice. Consider using this application guide collaboratively with colleagues to anchor your exploration, discussions, and application of your learning.

Chapter 1 Application Guide

Chapter Themes and Tools	Connect and Apply to Your Practice
Avoiding tug-of-war	Consider your students. What types of interactions feel like tug-of-war? What are the emotional undertones of these interactions? What do you need to do to help you "put down your side of the rope"?
Identifying and unhooking your if-thens	Jot down some of your if-thens. They can relate to student behavior or something else. If holding on to a particular if-then isn't doing you any good, try to unhook it by taking a new perspective.
Taking a new perspective	What do you notice when you try taking perspective to unhook an if-then? What part of that process comes naturally to you? What part takes a little more work?

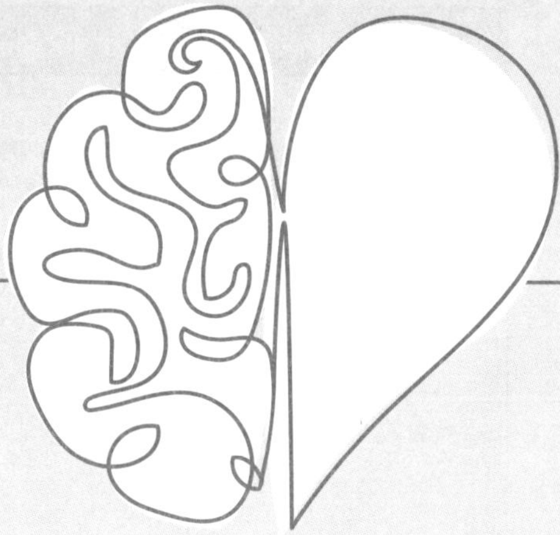

I see you,

I hear you,

and what you say

matters to me.

—Oprah Winfrey

CHAPTER 2

When Am I Ever Going to Use This?

It's sixth period and the end of an unusually hot, late-spring day. Only twenty-three minutes separate us from the weekend. This is a tough time of day for students in mathematics intervention. After exerting so much mental and social energy throughout the day, I ask them to learn content that's historically particularly challenging for them. Depending on how the day has gone, as teachers, we might find ourselves running on near-empty reserves of patience while the inevitable student shenanigans and varying needs for academic support remain.

"Teacher, when are we gonna *use* this?" Alexei interjects as we transition from a whole-class discussion to small-group work. *Oh, Alexei,* I think. I wonder if students can read the hints of exasperation on my face. *Breathe, Jennifer.*

"Bro," another student in Alexei's group replies while gesturing at me, "you can just tell her if you need help."

I try to maintain my poker face, but inside, I'm beaming a bit with pride as I witness the outcome of my efforts to create an emotionally and intellectually safe learning environment. Alexei's groupmate had been in my class the previous semester and was well-accustomed to my typical response to this question. It is rewarding to hear students embrace and reiterate messages we work hard to consistently convey. In this instance, I heard echoing back to me, *It's OK to ask for what you need. We are all here to help.*

"Nah, I don't get it. I don't know what we're doing," Alexei states as I approach his table.

Notice how his honest self-evaluation in this moment is so different from his initial question. Perhaps there was something inherently combative about the original question that put me a bit on defense, feeling the need to explain *yet again* that engaging in learning mathematics is important. Maybe it feels futile because we know from experience how little our specific, accurate answers will matter; after all, that is often *not* what our students are really asking.

"We can figure it out together," I say, making the effort to keep things relaxed. I take on a support role and volley responsibility back to Alexei and his groupmates in a move that aims to create shared engagement in the learning.

"What are the directions for this task?" I stick with Alexei's group and ask a few more questions to ensure they are headed in the right direction. Then I move on, checking in on the remaining groups to ensure similar confusion is not hindering others.

Choosing an approach to re-engage students with mathematics and nurture their sense of belonging in our learning community requires quickly discerning what factors may be at play and targeting a response accordingly. To understand the nuance nested within any given classroom moment, we need to slow down our inquiry and develop our skills to analyze students' mathematical and personal backgrounds. Let's use the interaction with Alexei as a case study.

Getting to Know Alexei

Allow me to introduce you to Alexei. An eighth grader, Alexei was enrolled in second-semester mathematics intervention. He had a cunning sense of humor and a readiness to spar tucked just behind his cool and collected exterior. I vividly remember two interactions with Alexei, both of which occurred during the first week of the new semester. I share them here because they add dimension and context to the circumstances surrounding Alexei's question, and they paint a more complete picture of the dynamic he brought with him to class. Attending to details, as minor as they may seem, is one key decoding skill you can develop to better understand the messages students are actually trying to communicate (despite the words they are choosing).

In *Culturally Responsive Teaching and the Brain*, author Zaretta Hammond (2015) shares "Culturally Responsive Brain Rules" (p. 46). Her first rule states, "The brain seeks to minimize social threats and maximize opportunities to connect with others in community" (Hammond, 2015, p. 47). To that end, we must invest time in community building with any group of students, especially groups with a mix of new and returning students, due to the risk of exacerbating any outsider feelings. Intentionally cultivating togetherness minimizes social threats and fosters a sense of belonging, which allows our brains to engage in more impactful learning (Hammond, 2015).

My first memorable interaction with Alexei transpired during a class warm-up activity. In this particular community-building activity, we were sharing about our hometowns. (Remember the curiosity students often expressed about my origins?) I was acclimating to the new class dynamics, connecting with former students, and meeting those new to our class. When asked about his hometown, Alexei named a major city in his home country and asked if I knew where it was. I responded enthusiastically, pleased to have some geographical point of reference for this student. I knew he was new to Oregon, and I was eager to hear more about his life experiences prior to moving here. "I haven't been there, but I know where it is!"

"My hometown's *not* near there, teacher. You drive and drive until you are almost dead, and you are still not in my town." His classmates laughed at his dry, effortless humor, and so did I.

"Not near there. Got it. Thanks, Alexei. Welcome—I'm glad you're here!"

Whether in that moment or in reflection after an exchange as quick as this, practice slowing down the interaction and considering the nuance as you practice picking up on the embedded belonging-themed signals. The subtext to that discussion might have been Alexei seeking to know: *What is this teacher's scope of knowledge about other places? Do I have any shared experiences with others who know? Is there any hostility or familiarity among the class? Is it safe to share something about what is unique to me? Do I belong here?*

The second memorable encounter with Alexei took place later that same week. Students were lined up at the door just moments before the bell rang. We were chatting informally about their plans for the upcoming weekend.

"Alexei, what about you? Any plans for the weekend?" I asked, trying to bring him into the discussion.

"Dunno," he said quietly as he shrugged and looked at his feet. "Maybe get in a fight."

I'm not sure if he said this because he trusted me or was going for shock value. I tried to sound curious rather than disapproving in my reply. "A fight?" I inquired. "Fight about what?"

"Don't know yet. Just a fight."

"That doesn't sound like very much fun," I said, trying to be nonchalant yet also listening for details that may be important.

"Not much fun for the *other guy*," Alexei continued. "But fun for me."

All right, I thought to myself. I decided this fight did not sound like a planned event that needed my intervention, so I let his comment be. I guessed next week we would find out if, in fact, fights had been part of Alexei's weekend. Again, with this interaction, slow down to examine the subtext: *What are the undertones about a physical fight? What are the social consequences for sharing this in class? With a teacher? What might Alexei be communicating about his sense of belonging outside our classroom?*

These two interactions served as my initial data points about Alexei. Of course, these moments did not represent the *totality* of Alexei's characteristics or complexities, but they certainly helped provide context for the person seeking to engage in the moment he asked, "When are we gonna use this?"

What's Really Going on Here

If you have taught mathematics for longer than three class periods, you have likely heard the infamous, "When am I ever going to use this?" I remember student teaching and observing expert teachers provide compelling arguments in response to students who posed this question in class. It seemed they could relate students' current or future lives to the mathematics they were learning in nearly every instance. Yet, I noticed students smirk to one another or nod along sarcastically as the teacher responded. The nonverbal cues in this exchange brought up the question of what *really* prompted students to ask that question in the first place. Given that teachers' answers rarely seemed to satisfy whatever need was *beneath* the surface of the question, perhaps an alternate interpretation was needed.

PAUSE AND PONDER

What does your experience tell you about what is at the heart of the question, *When am I ever going to use this?*

There are at least two interconnected threads tangled together in the question, *When am I ever going to use this?* One is the *human connection* (belonging) thread—students expressing their sense of disconnection from other people. The second is an *access to learning* (engagement) thread—students expressing their disconnection from the content.

Visualizing these as tangled, intertwined threads can help us understand how a student may come to us holding one of those threads and essentially communicating, "This one has a knot." If you've ever untangled or unknotted anything, you know that to free that single strand, you need to free all the other strands it's connected to. Students show you just one strand of their current experience, and it's important to convey the following in return: *I see the thread you are holding, and I know it is connected to other things. The knot might be in that thread, or it may be someplace else. Either way, we will find it, and we will untangle it. Knots can be a lot to manage and untangle. I'm here to help.*

Untangling the Threads

The first and crucial step is to gain some clarity from students about what they really need. As redundant as it can be to hear the question, *When am I ever going to use this?* imagine the lack of belonging (or *outsider-ness*) students feel when disconnected from the people or content they are learning.

Depending on the moment and the dynamic, it may be best to ask your follow-up questions off-stage, directly with your student. Just because a student asked you a question in front of the whole class does not mean you must respond in front of the whole class. Take a breath and a few seconds to scan the room to get a read on the *class tone*—the complex interpersonal dynamics among students. This moment to slow down the exchange will improve the likelihood of successfully deciding whether this is a whole-class moment or single-student moment. Since we are not in the business of outsmarting or one-upping our students, we can take an earnest approach to uncovering what students each really need, modeling the curiosity that signals the importance of each student in your class.

If you are asked the question in a whole-class setting, but have the sense that it would be best answered one-to-one, consider saying something like, "I hear you. Hold tight, and I'll answer you in just a moment." This response honors the student by acknowledging you heard the question and simultaneously provides the time you need to get the class engaged in small-group discussions so you can focus your attention on that student.

Once you can converse one-to-one, you might ask a few questions to better understand what is really going on, seeking to identify which threads are tangled up. Here are a few clarity-seeking questions to ask.

- "How do you see this related to your real life?"
- "Do you need help getting started?"
- "Is this confusing?"
- "Are you feeling frustrated?"
- "Do you need my attention?"

You may be wondering, "Why do my students ask a 'cover question' instead of what they really need?" There may be several reasons. One significant consideration is students risk much more vulnerability and exposure when answering the questions in the preceding list, compared to when they ask, "When am I ever going to use this?" Think about it: this question essentially asks for a head-centered response, a logical or well-reasoned justification. The alternate interpretations—such as the possibility of needing help or feeling confused or disconnected—are far more heart-centered and vulnerable needs. It can be a high-stakes gamble for students to voice these needs if they are unsure of the safety of their learning environments. The following sections provide some ways to craft head and heart–type responses to the questions students are really asking.

Being Culturally Responsive

As is often the case when we learn and grow as professionals, I found the words and understanding to contextualize my teaching experiences later in my career when I dove into the work of Zaretta Hammond (2015). In an endeavor to untangle the *human* and *content* connections our students need, it is crucial to lean into the *cultural responsiveness* Hammond (2015) describes, which means that understanding and honoring the context students bring to any given mathematics problem (how they think about, solve, and engage with mathematics) depends on their lens, culture, and context.

In her list of actions to facilitate "information processing" (content connection), Hammond (2015) calls us to "connect new content to culturally relevant examples and metaphors from students' community and everyday lives [and] provide students authentic opportunities to process content" (p. 17). In our profession, we frequently hear that if mathematics is put in a *relevant* context, students will be more likely to engage. Being culturally responsive is just one of the two key factors to facilitate information processing, however. In the Providing Accessible Tasks section (page 35), we'll explore the second facet, which is how to provide the authentic opportunities to process the content.

CONNECT EARLY AND OFTEN

When your tangled-thread analysis reveals a knot in the human-connection thread, the approach and response are different from when the knot is in the content-connection thread. Culturally responsive educators forge "learning partnerships" (human connections) that "give students language to talk about their learning moves [and] reimagine the student and teacher relationship as a partnership" (Hammond, 2015, p. 17). Every so often, *When am I ever going to use this?* is more accurately translated as, *Will you pay attention to me?* Consider the possibility a particular student who asks this of you *every day* is wondering, *Do you see me?* For this student, investing in a learning partnership may mean greeting the student at the door and pulling the student aside with kindness—as if you are having a top-secret conference about mathematics. With a look of enthusiasm and anticipation, say something like, "Hey, Myra, today we're working on percentages, and at some point, we'll talk about where we see percentages in the world around us. Be thinking about that so when we get to it, you can start off your group's conversation."

This is often all it takes to set the stage for engagement and belonging—by conveying, "I see you, you matter, and your contributions to this class matter." You may even skip the mathematics connection altogether and greet or connect with students as they're walking in to provide a touchpoint, particularly for the students you know will appreciate it.

As you likely know, not all students prefer to have the attention of the teacher in this way. Certainly, at the middle and high school levels, plenty of students would do anything *not* to have the teacher's attention. But a few students need additional or more frequent connection points to feel like they belong. Additionally, this strategic move preemptively makes space for student voice and cultivates the social and intellectual safety necessary for the brain to learn (Hammond, 2015).

In other instances, you may notice a pattern with students interrupting instruction or group work as they seek to connect with their peers or you. Students leaving their groups to ask, "Did I get this right?" may be asking about the mathematics, or may, in fact, be asking for your attention—a bid for connection. Gather some data over a period of time, and notice what patterns emerge. Stay curious about the motivations for different behaviors, and as you nurture those learning partnerships, students will grow more comfortable speaking about their under-the-surface needs.

MAKE REAL-WORLD CONNECTIONS

It is possible your students *do* want to know the practical application of the mathematics they are learning. In that case, consider volleying the question back to a student or the whole class: "When *might* we use the mathematics we are learning today? How *is* this connected to something you already know?" This is a rich opportunity for students to seek and make connections to the real world and their existing funds of knowledge (Hammond, 2015), and for teachers to learn about how students see relationships between mathematics and their lives.

Alternatively, consider preempting the question by starting a lesson with, "Today, we are talking all about percentages. Where have you seen or used percentages in *your* lives?" Since we know the importance of helping students make connections to mathematics through their real-life experiences (Arthur, Owusu, Asiedu-Addo, & Arhin, 2018; Matthews, 2018), why wait for them to ask? Be ready to fill in the connection gaps by using guiding questions as an integral aspect of lesson planning that sets the stage for engaging students throughout your lessons.

Sometimes, the actual mathematics students are learning does not have a compelling, direct, real-life application. When in real life do we need to know how to factor quadratics? Rarely, depending, of course, on your career. When in real life do we need to know how to isolate variables, solve for an unknown, work collaboratively, make a prediction, test a theory, or adjust a strategy? Almost always!

It can be helpful to engage students in a whole-class discussion about the non-mathematics topics they are learning alongside mathematics topics. Consider the power of stories to shape the examples you share with students (such as how some runners cross-train by swimming, soccer players by running, gymnasts by lifting weights, and swimmers by doing on-land training). Likewise, as your students share stories of their lives (such as conflicts with peers, learning an instrument, playing sports, debating with their home adults, and noticing patterns), make connections to how they are engaging in the world in mathematical ways. The same habits of mind they use outside class are applicable and needed during class. Discuss these habits of mind, the perseverance students are practicing, and how, although mathematics might not be their future "main event," it is important cross-training for whatever their main event might be. In those discussions, continually reinforce with your students that mathematics belongs in their lives, and that they belong in the classroom to study mathematics.

PAUSE AND PONDER

How do you get to know your students to ensure you provide mathematics learning contexts that are relevant to them?

Providing Accessible Tasks

If you discern the source of the tangle is a content-connection knot, the most impactful instructional move is to consider the task's *accessibility*. Making changes to lesson plans and adjusting on the fly to improve accessibility untangles a good portion of frustration and re-engages students in learning. As I practiced my interpretation skills, I came to hear the cover question more accurately translated as, "I'm pretty sure even if I try, I still won't get it." In other words, "I don't have access to this learning." We know, however, that an intellectual understanding of how mathematics pertains to real life does not meet the need here, because what students need but are not asking for is a way to get into the mathematics task (Fernandes, Crespo, & Civil, 2017).

Empathize with students who feel disengaged after repeated experiences trying to learn mathematics and repeated experiences of it not making sense. It's understandable for students' internal narratives to sound something like, "If *this* mathematics class is anything like my *other* mathematics class, I'm cooked." Disrupting that pattern, fostering engagement, and nurturing a sense of belonging requires students to encounter mathematics tasks with 100 percent guaranteed access. No matter what content standards you plan to teach, your students *cannot* achieve understanding without initial access. Accessible learning opportunities often include one or more of the following features.

- There is more than one way to think about the problem.
- There is more than one way to solve the problem.
- There is more than one *acceptable* way to represent the solution.
- Engaging with the task includes collaborating with others.

- Tools or manipulatives are readily available.
- Students do not need to depend solely on the teacher to get unstuck.

Notice some of the elements in the preceding list relate to the task itself and some relate to how to facilitate a task, that critical aspect of "providing students authentic opportunities to process content" (Hammond, 2015, p. 17). The tasks you choose and how you facilitate them go hand in hand to create accessible learning opportunities. Depending on which authors or experts you are familiar with, you may hear about *open* tasks or tasks with a *low floor, high ceiling* (Boaler, 2008). You may even hear them use similar features to describe *rigorous* tasks (Marzano, 2019). While these terms are not synonymous, they all describe tasks that include access points for all students (National Council of Teachers of Mathematics [NCTM], 2014). For trusted resources about accessible tasks, check out appendices A (page 155) and B (page 161).

By contrast, tasks with one single entry point (one "right" or acceptable approach or one acceptable way to represent an answer) feel like narrow ledges for students to traverse. Instead of embarking on a risky, high-stakes journey, students will opt out, deflecting attention with questions like, "When am I ever going to use this?" It is not exactly an *off-task* question, but when students ask it in this context, it certainly isn't *on-task* either. When you hear it, consider that perhaps there is an accessibility concern requiring your attention.

Aside from the task itself, the way you facilitate it or the learning structure you use also impacts the task's accessibility. Four key elements of accessible facilitation are (1) representations, (2) manipulatives, (3) discussions, and (4) independence.

CONNECT REPRESENTATIONS

Often, students engage in mathematical representations one at a time. For example, they interpret data in tables in one lesson, graphs in another lesson, and equations in yet another lesson. Instead, structure learning experiences that engage students in processing content using multiple representations during the same lesson. Normalizing and expecting students to use multiple representations of their thinking (words, pictures, tables, graphs, and equations) simultaneously within a single lesson is a facilitation move that creates accessibility, cultivates engagement, and promotes deep learning (Hattie, Fisher, & Frey, 2017).

Teachers don't expect *some* students to draw pictures, and *some* students to write equations. Teachers expect *all students* to use and connect various representations and identify the mathematics throughout. Additionally, plan discussions that surface the connections between different representations, and invite students to share which types of representations they prefer (or find most useful) in different problem contexts. This may be challenging for students initially, but it is also a sign their brains are growing their intellective capacity (Hammond, 2015).

USE MANIPULATIVES

Secondary mathematics classrooms utilize manipulatives occasionally (or even rarely) in comparison to most elementary classrooms. However, we know how crucial it is to student engagement and deep learning to use manipulatives throughout learning to "make concepts concrete and visible, look for patterns, make connections, and form generalizations" (Hattie et al., 2017, p. 170). Select manipulatives that align to the content and ensure they are within reach on student desks, eliminating the potential shame or intimidation students might experience if they must leave their desk to retrieve a manipulative.

PAUSE AND PONDER

What other ways might our classroom routines or arrangements unintentionally instill a sense of shame in students?

How might shame impact students' sense of belonging?

Make it clear to your students that using manipulatives is a sign of trying to figure something out, not of any lesser or underdeveloped intelligence. Facilitate ongoing discussions, progressing from physical representations with concrete tools to visual representations and, eventually, symbolic representations. Students have shared that it felt like the "smart kids" did not need manipulatives; they could just write the equations. We must help students unlearn this notion by structuring learning opportunities in which the teacher expects a student who can write a linear equation to explain how it relates to a growing pattern displayed in color tiles, just as a student who can build the growing pattern learns to write symbolic equations. Since students need deep, connected opportunities to engage in learning, they can all benefit from using manipulatives. To encourage manipulatives use universally, keep them within students' reach.

SUPPORT RICH DISCUSSIONS

All mathematics-intervention students (not just those learning English as a second or third language) need support with understanding and producing mathematical language. Sentence frames are one support you can use to foster mathematically productive discussions. If you are new to the idea of using sentence frames in mathematics class, I encourage you to jump in with both feet. They change everything when it comes to students accessing the words they need to capture mathematical ideas, communicate with peers, understand one another, and ultimately engage with and understand mathematics (Melanese, Chung, & Forbes, 2011).

In addition to using sentence frames, another critical component of facilitating rich discussions is to get out of the way. Say less. Be patient. Stay curious. Ask more questions. Let students wrestle with the content and debate their ideas with one another. Provide ample time for discussions and the process necessary for making sense of mathematics, and guiding through questions, and avoid "rescuing" them when students show confusion. Investing time up front in making mathematics discussions meaningful will enhance student learning (Hattie et al., 2017) and mitigate the disconnection they predictably and cyclically feel when they do not get to develop deep understanding (Hammond, 2015).

FOSTER INDEPENDENCE

Creating classroom learning environments that push students to develop self-efficacy and student ownership of learning is an aspect of accessibility that calls for some amount of independence from the teacher (Hammond, 2015). Since there is only one teacher (typically) in class, it doesn't work for students to rely solely on the teacher to help them get unstuck. Not only does it not work in the moment, it also doesn't work long term when we think of the problem solvers and mathematically confident students we're hoping will move on to their next endeavors.

As you plan lessons and consider how best to cultivate engagement, foster independence using systems students can replicate in any aspect of their lives they are self-managing. One of my special education colleagues showed me her *sticky note system*, which she uses to support students with keeping track of multistep processes, and I immediately found value in teaching all my students the approach. The general idea is that students write each step of a process on a sticky note and place the notes vertically in order of the process on the left side of their desk. (I chose the left side of the desk as the starting place because we read from left to right.) Then, as students complete a step in the process, they move their sticky note from the left side of their desks to the right side. This helps students slow their process, develop the *metacognition* (the ability to think about their own thinking) required to self-evaluate their progress, and understand which aspects of a task they may be struggling with.

"It is not an accident that students become more engaged in their own learning when they are actively involved in tracking their own progress . . . focus[ing] attention . . . cause[s] the brain to rewire itself and grow" (Hammond, 2015, p. 101). The sticky note system can help conversations move from "I don't get it" to "I understand ___________, but I'm confused about ___________." In addition to the metacognition students develop, moving the sticky notes becomes intermittent indications of success—or *mini-finish lines*—plus a valuable formative assessment tool for teachers. Even if students struggle midway through a problem, there is a sense of achievement with each sub-step that bolsters the motivation students need for sustained independent or small-group work.

It is important to note this type of system is not intended to facilitate memorizing algorithms. If, for example, students are working on dividing fractions, I would not ask them to create a sticky note set for a procedural mathematics "trick" (for example, *keep-change-flip*). This shortcut, although teachers commonly use it to help simplify an operation such as how to divide fractions, is in effect, a process devoid of meaning for most students. Most students cannot tell you *why* a tool like keep-change-flip works. Memorizing shortcuts and tricks often disrupts sensemaking, which is what my students really need the opportunity to do—develop conceptual understandings and internalize mathematics. Instead, the sticky note system for working with fraction operations might include the following steps to scaffold building conceptual understanding.

1. Draw a picture of the fractions in the problem.
2. Restate the problem (for example $\frac{3}{4} \div \frac{1}{2}$ is asking me how many groups of $\frac{1}{2}$ there are in $\frac{3}{4}$).
3. Show the grouping in my picture or with my tools (using fraction tiles or fraction pieces).
4. Check with a partner.

There may be several more threads, in addition to human or content connections, you discover as you untangle the question, *When am I ever going to use this?* As you endeavor to cultivate engagement and belonging, remember to stay curious about the source of the tangle before planning your response or instructional approach. Otherwise, you may spend an inordinate amount of time answering questions no one is really asking, a frustrating experience for all parties involved. Bring some extra patience with you on this journey (it can take time to develop your skills for uncovering the real questions students are asking), create a repertoire of accessible mathematics tasks, and refine your facilitation skills.

PAUSE AND PONDER

How do you ensure a task is accessible?

What additional ideas do you have to help students become independent learners?

To Sum It Up

Discern what students are trying to communicate and answer the question they're not yet asking.

"But *teacher*," Alexei persists on a different day, during a different lesson. "I'm never going to *need* to know this!" There seems to be a collective exhale as instruction pauses. While I may initially interpret his question as off-track, perhaps intentionally disruptive, a momentary pause to notice the tone in the room clues me in on the fact he was likely advocating for more than just himself when he interjected his question, effectively pressing pause for everyone.

"Hold on," I say slowly as I survey the room and look more closely at students' faces. "Are y'all feeling frustrated?" I genuinely inquire.

"Yes! Duh!" they declare in unison.

"My bad, team! Alexei—thank you. I didn't catch it 'til you spoke up. How about we take a two-minute break? You can play the games at the back table, stretch, chat, hang out. I need a second to figure out what's next." I give myself the space I need to consider the students' feedback. I hadn't realized their tension levels were inching up. I shift gears, thinking more time for discussion and wrestling with the current mathematical ideas can wait until tomorrow. Instead, I want to return to content with which students feel more at ease. My hope is that I can prevent a mutiny from arising out of the simmering tensions.

"All right, I have an idea. But I need your help knowing where this all started to get confusing. So, walk it back with me." Students return to their small groups, and I rewind the lesson until we find the tension point where much confusion started to arise. Given the events of the day,

I decide to mentally "pin it" at that tension point and pick up the content the next time we meet.

"Thanks, team. I think I understand where this started to get really confusing. Using our sentence frames on the wall, let's pause and wrap up this learning in our journals. Write three sentences about your learning today and be prepared to share with a partner."

You can imagine how the remainder of the class played out. Students wrote, then they read what they wrote or had an open discussion with a partner. I circulated and listened, gathering important formative assessment data about our learning needs. Once students put their journals away, we shifted to a series of games focused on building number sense. These games were familiar to students, they enjoyed playing them, and they provided the change needed to prevent pointless struggle from being the primary experience of that class period.

I knew when I first met Alexei that it might take weeks of listening, asking for clarification, and connecting before I'd see a shift in his learning patterns. Nonetheless, I kept at it, persistent in my efforts until he, too, started to vocalize his needs more accurately. Whether you teach at a large or small school, in a rural or urban setting, many who cross the threshold into your classroom need opportunities for human connection as much as they need academic support. As students are developing their multiple, complex identities, this approach to meeting their needs for content and human connection contributes to the general sense that students belong in each learning moment, and each learning moment belongs to them.

Chapter 2 Application Guide

Use the reproducible "Chapter 2 Application Guide" (page 42) to gather your thoughts and make connections between the themes and tools in chapter 2 and your teaching practice. Consider using this application guide collaboratively with colleagues to anchor your exploration, discussions, and application of your learning.

Chapter 2 Application Guide

Chapter Themes and Tools	Connect and Apply to Your Practice
Untangling the threads	What clarifying questions will help you better understand what students need? When will you find time to engage a student in one-to-one discussion to gain clarity?
Being culturally responsive	Consider your students' backgrounds and experiences for an upcoming lesson or chapter. Do you anticipate they'll be familiar with the content? If so, how can you leverage their expertise throughout the lesson? If not, how might you bridge the gap between their experiences and the mathematical content?

Connecting early and often	How might you support the varying degrees of interpersonal connections students need?
Providing accessible tasks	Look ahead at a lesson, chapter, or unit of instruction and identify the four elements: (1) representations, (2) manipulatives, (3) discussions, and (4) independence. Do you need to make any modifications or supplement the content to incorporate the four elements? Start with one small change. What will that change be? How will you know if the change helped students access or stay engaged with learning?

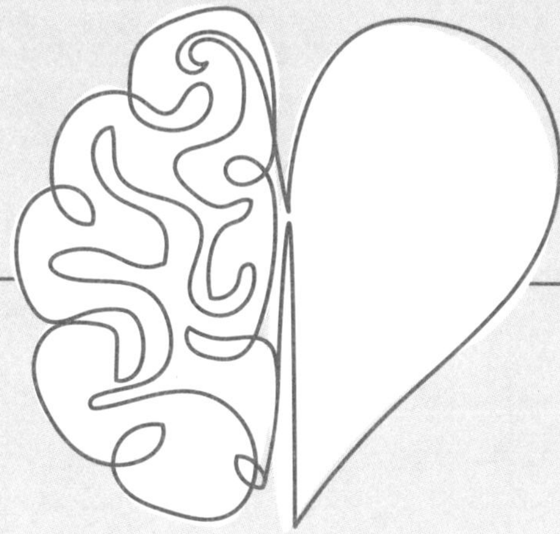

You never forget

who sees you first.

—Steve Pemberton

CHAPTER 3

Not Today, Ms. Lenhardt

It is the beginning of the semester in my mixed-grades mathematics intervention class, and I'm sitting at my desk while my students finish their preassessments. I feel confident and excited. I have my thoughts on curriculum, lessons, units, individualized educational plans (IEPs), and English proficiency plans, and I am *ready*. As students finish their preassessments, they walk to my desk and hand them in, then they begin working on a writing prompt while their classmates finish up. (Yes, we were writing in mathematics. Like I said, I am crushing it by incorporating it *all*: A+ teacher planning!)

Up walks Diego, a student in the seventh grade I had yet to learn much about. He looks me directly in the eye as he slams his preassessment on my desk. A curse word (starts with *f*) and the word *you* are scrawled across the page, all in capital letters and underlined several times. This emphasis is, of course, immensely helpful. Without the underlines, I am not sure I would have gotten the message.

PAUSE AND PONDER

How are you feeling as you read this?

What would you do in this situation? (Or what have you done?)

I have a micromoment to decide how to respond and react. What seems like ten zillion questions race through my mind.

- Do I enforce the school rules and follow discipline procedures according to the student code of conduct?
- Do I categorize him in my mind as a disrespectful student?
- Am I offended?
- Should I revisit our school's *Be respectful* mantra?
- Am I fearful, angry, disappointed, annoyed?

There are so many things to consider. I didn't have the answers to every question racing through my mind, but I did have a strong instinct not to escalate the situation. Here is what unfolded next.

"Got it," I said as he emphatically set his preassessment on my desk.

"That's it?" he said, somewhat incredulously.

"Yeah, that's it," I replied, looking right back at him with my absolute best neutral face.

"You're not going to write me a referral?" he inquired. "For *that*?" He gestured at his paper.

"Do you *want* a referral?"

"Well—no," he stammered. "I just always get in trouble, so I figured this would be the same."

"If you're in detention, how are you gonna learn any math?" I asked with a sideways grin I hoped conveyed I was not angry, nor was I intimidated.

"Ha! I *hate* math," he said. While his words remained angry, his posture visibly softened.

"Hmm, so I see. Thanks for the heads-up," I said as he grabbed his journal and lingered at the edge of my desk. "We've known each other for, like, two class periods or something, right? So how about we figure out what is so infuriating about learning math? It will not be awful forever, I promise. Thanks for letting me know how you feel."

I watched as this young student turned around and headed back to his desk, opened his journal, and started to draw. This was the first step toward building a relationship with Diego that would be crucial to his learning successes over the two years he and I would share learning space. The students nearby pretended not to be eavesdropping on our conversation, but I'm sure they were conducting their own sort of assessment about me, learning what sort of class this was, and deciding whether this is a place where they might find belonging.

These sort of confrontations with secondary students can be frequent and easily spiral out of control. In examining the dynamics at play, we can deepen our understanding and then tactfully and strategically de-escalate —even restore connection—to continually reinvest in creating spaces where engagement and belonging can thrive. Let's begin by considering the depth and nuance of this situation.

What's Really Going on Here

I didn't construe Diego's anger or what he wrote on his paper as disrespectful. I remember feeling a deep, connected knowingness. I know this feeling so well, and I am intimately familiar with the traditionally unhelpful educational spaces that fall short of teaching us what to do with all that anger. I was a kid who could get really angry, really fast. Like blood-boiling-in-my-toes kind of angry. My anger was frequently met with, "You have no reason to be mad." Has this ever happened to you? Someone says to you that you are feeling the wrong thing? Having another person dismiss your experiences and emotions is really helpful, right? *No*, fine people of the jury, it is in fact, *not* helpful.

What most of us need when we are in the heat of any emotion (but especially *hot anger*) is a tool or strategy to help us know what to *do* with that emotion so the feeling doesn't "run the show." And we often need the support of another person to help facilitate a return to our less-upset selves. It is never helpful to be told a feeling is irrelevant or unnecessary. That is the lesson I hoped to convey to Diego on one of our earliest opportunities to get to know one another. "It's OK to be mad. It's even OK to curse. I don't have any idea what went down in your day, but it's OK to be mad."

You may be wondering, "Didn't you just show all your students they can be disrespectful and get away with it?" Perhaps. This certainly was the concern behind my colleagues' well-intentioned advice. "You have to demand respect, or these kids will destroy your class," I remember more than one person warning. This may strike you, as it does me, as a troublesome perspective and a potentially unnecessary pitting of students against teachers. I took a calculated risk, at that time and place, as I let the events unfold in an effort to better understand my role in supporting individual students and the entire class in navigating their emotional landscapes.

What guided my interaction with Diego was a hope that I would convey to him I saw *him*, not just *his behavior*. I hoped his classmates picked up the cues: *This class is a place where big emotions are normal, where we can learn to do better, and punishment and consequences are not the "currency" for learning.* If we have any hope of actually impacting academic learning, students with the toughest exteriors and the tenderest spirits need to know we prioritize compassion, listening, and power-sharing (Boaler, 2008; Howard, 2019). And most importantly, do the work; learn how to notice and attend to your own emotional responses so you can be present and make plenty of space for students' feelings and experiences.

This interaction occurred so early in my getting to know Diego that I really was not sure what it all meant at the time. My instinct was to give him a referral, but doing so would not only have been a premature response but also perpetuate some of what led Diego to mathematics intervention in the first place. As many teachers do, I had an organic interpretation of our school's discipline policy, which meant discerning what I could best leverage as a guideline rather than a mandate. While I held zero tolerance for weapons, I simultaneously held lots of tolerance for cursing.

In many instances, we can choose *creative insubordination*, a term Gutiérrez (2016) describes as including "decentering the achievement gap; questioning the forms of mathematics presented in school; highlighting the humanity and uncertainty of mathematics; positioning students as authors of mathematics; and challenging deficit narratives about students of color" (p. 54). Rather than enforcing consequences that do not serve individual or collective student needs, we must choose approaches that reconnect students with their own curiosity and love of learning (Gutiérrez, 2016). Students must encounter *restorative*, not *punitive*, schooling experiences (Darling-Hammond, 2023). How will these students ever have a shot at learning anything if teachers perpetually cite them for every infraction and hold them accountable for every less-than-perfect moment? The following section digs into some ways we can approach this.

Avoiding Zero-Sum Games

The power dynamics among teachers, administrators, staff, and students require careful attention and intentional decision making if we hope to continuously disrupt harmful cycles of *zero-sum games*, a mindset or system of beliefs that says if you gain a point (+1), then I lose a point (-1; von Neumann & Morgenstern, 1944/2007). Your +1 combined with my -1 results in a sum of zero, thus the term, *zero-sum game*. Although I first learned about this concept from a discussion of power dynamics between nation states in a political science class, I quickly found this framework useful in understanding the politics of my classroom and the power dynamics between myself and students (Gutiérrez, 2013).

Consider a student interaction like this through a zero-sum lens. A teacher and student who are locked into a zero-sum framework respond to each move from the other party from a -1 standpoint. If the student is up 1, then the teacher is down 1. In the zero-sum construct, you must escalate your approach in a crush-or-be-crushed fashion to win the interaction. We have all seen this unfold in real life. Repeated instances of these power dynamics and zero-sum interactions have a cumulatively devastating impact on both teacher and student. The term for the final outcome of zero-sum games is *mutually assured destruction*. It may sound dramatic, but it's what happens when we operate in this framework. This is just not the cycle we want to be in.

The great news is it does not have to be this way. By *not* escalating your response, you issue an invitation to play the learning and power game by a different set of rules. You show the student (and those watching) you're not there to "win" by overpowering anyone. People are deeply conditioned to see conflict or tension in a zero-sum framework, and it can be challenging to see another way or operate from a different perspective. I am not saying it's *easy*. I am saying it's *necessary*. Here is some guidance for stepping out of the zero-sum construct.

- **Learn your hot spots:** Study your own motivations and experiences to understand what activates your zero-sum mentality. It helps to learn to see it coming so you can choose a different response (Hammond, 2015).
- **Gather data:** If behaviors are the "tree," we want to know more about the "roots."
- **Create a power-sharing dynamic:** Reframing power dynamics must start with those who traditionally hold the power.

Let's examine what each of these steps entails and how this process can empower you to slow down the "rage train."

LEARN YOUR HOT SPOTS

Introspective, important work must be done to better understand our own experiences and *hot spots*, the things that activate our "fight-flight-freeze response" (Siegel & Bryson, 2011, p. 129). This work is necessary to learn new ways of interacting with students and step away from zero-sum games. Research and practice support this notion. Daniel J. Siegel and Tina Payne Bryson (2011), coauthors of *The Whole-Brain Child*, describe employing a "name it to tame it" strategy (p. 27), and Hammond (2015) writes about the importance of "identify[ing] your triggers" (p. 62) and "labeling your feelings when they come up" (p. 67) as critical parts of preparing to be a culturally responsive practitioner.

Creating a mental model or picture of what this research means can help you understand the implications. Try visualizing dominoes standing up on end, each domino representing a memory or life experience, carefully standing up in a winding serpentine through your years on this planet. Being activated is like having that first domino tipped over. Before we know it, the tumbling of all subsequent dominoes feels inevitable and out of your control.

PAUSE AND PONDER

How do you know when something activates you? What happens for you emotionally, physically, and mentally?

What happens next can look like experiencing emotions that outweigh the single initiating event or mentally pull up a string of historical events. You may feel as if you're reliving, rather than merely remembering, the emotions accompanying each event as if it's happening all over again. This may look different for each of us. I know in my case, the "tumbling of dominoes" typically manifests initially as rage and blame. (*This wouldn't be happening if* __________ *had not* __________.) This morphs into an extraordinary number of tears, representing unhealed wounds. Because we bring our beautifully complicated selves to teaching, we are responsible for learning how to recognize when our "dominoes are tumbling," and take steps to care for ourselves and regulate our responses.

When you feel activated, you know you need to get back to some sort of emotional equilibrium before you can confidently respond (instead of react) to a situation. The ongoing emotional development work is about learning how to make space between the first ten dominoes and the next ten, or creating a barrier that prevents these dominoes from tumbling into the next dominoes. How do you do that? Paying attention to physical responses like heart rate, sweaty palms, and ringing ears will help you recognize when you're activated (Hammond, 2015). I timed myself once to see how long those physical symptoms last. It was about eight minutes for me, which is data I use to give myself space. I give myself the grace of at least eight minutes after an activating event before I even begin *considering* my next move.

You *cannot* always control or prevent the activating event, that initial "flick of the first domino." What you *can* control is whether you mentally stand by and watch "all the dominoes fall," bewildered why this has happened *again*, or whether you place your proverbial hand (in my case, *eight minutes*; you could try deep breaths or brisk walks) right between "two dominoes" to stop the subsequent tumbling.

PAUSE AND PONDER

How do you, or might you, make space between your "dominoes"?

Identifying hot spots is personal, and complex, and can include the support of trusted friends, colleagues, or even a skilled therapist. I cannot endorse with enough emphasis the value of doing this work and finding professionals to help you if you need them. It is worthwhile work for all aspects of your life and particularly helpful as you encounter hundreds of students with their millions of "dominoes also tumbling" at unpredictable times for myriad reasons. This internal work is a crucial step in the process of securing mutually assured thriving, rather than the destruction that comes from zero-sum dynamics.

Remember, your students are still learning how to manage their "dominoes." It's part of growing up, and you hear it in raised voices, slammed doors, and so many other ways. Recognize fury and know that learning how to stop the "domino tumble" is a skill we must learn and practice. This perspective affords you a lot of patience in teaching the needed skills through de-escalating and restorative practices (Howard, 2019).

GATHER DATA

Formative assessment is a collection of strategies teachers use to gather information about student understanding to inform instruction (Joyner & Bright, 2016). In my teacher preparation and schooling, I recall an entire class (or at least a six-week stretch of a class) dedicated to learning approaches for formatively assessing students' mathematics knowledge. It did not occur to me until I was in the classroom that the same skill set I was developing for an academic assess-listen-respond cycle would be critical to the necessary process of building relationships with students.

In education, we talk a lot about formative assessment data, how to gather and what to do with the data. Typically, those conversations center on gathering information about our students' mathematical understandings. However, when it comes to re-engaging students in learning (especially in learning mathematics), we should leverage our formative assessment chops to gather information about our students' emotional status as well. We need to be well equipped to respond to data in a way that "moves the needle" on learning.

Students' emotional landscapes impact learning (Collaborative for Academic, Social, and Emotional Learning [CASEL], n.d.b; Darling-Hammond, Flook, Cook-Harvey, Barron, & Osher, 2020). A student's emotional state can render the most fabulous lesson plan completely ineffective. With each student interaction, whether affirming or challenging, questions to ponder include the following.

- What can I learn from a student's response that will inform how I approach *teaching this student well*?
- What are students showing or telling me they need?

Here are some useful approaches for gathering important follow-up data.

- Questions to ask students include the following.
  - "How did you feel about our lesson today?"
  - "What do you do when you are angry?"
  - "How can I help when you are feeling mad?"
  - "What do you need?"
  - "Can you help me know when you start feeling this way?"
  - "Tell me about another time you were this angry. What happened next?"

    + "If your friend came into class super angry, what would you say or do to help?"
- Ways students might respond include the following.
    + Writing in mathematics journals
    + Completing exit tickets
    + Drawing
    + Talking one-to-one
    + Opting not to respond

At this point, you may be thinking, "Listen, Jennifer. I do *not* have time to do this level of exploration about each of my students." You may be feeling uncertain and perhaps overwhelmed. I hear you. It can seem this time spent will interfere with instructional time. I assure you, my approach with Diego certainly benefited him as well as all the other students in the class not only in the short term but also (and most especially) in the long term. This investment builds the solid relationships necessary for learning to occur. What is the cost to try? Perhaps more importantly, what is the cost to *not* try?

PAUSE AND PONDER

What are some questions you might ask students to determine what they may need?

How might students respond to your questions?

It is hard to envision where to make time to gather this emotional formative assessment information. You can start small. Try adding one emotional check-in question to your exit ticket this week. Or try checking in with five students in some way, rather than every student. It can seem daunting, but this is the essential work of teaching we swapped out for other nonessential tasks (such as checking homework). We will explore more strategies for reprioritizing instructional time in chapter 7 (page 117) and chapter 8 (page 133).

The upshot of this approach with students is you can ask them a series of questions (in my case with Diego, a preassessment), and they will answer you with accurate, valuable information you can use to teach them well. Diego's response was incredibly informative for me as I tried to learn more about his mathematics learning experiences that led him to mathematics intervention. I could take offense to the language he chose, and

I may not understand why his response was so heated, but make no mistake, he had answered the question. The next part is up to us to determine what to do with the answers our students give. For students in situations like Diego's, the language they choose isn't a red flag; it is a burning SOS! What we know about formative assessment impacts our approach to this interaction and several other teaching interactions that require getting at the root of what we are observing in student behaviors.

CREATE A POWER-SHARING DYNAMIC

Teachers are, by default position and title, the *power brokers* in the classroom. This means it is our job to distribute power intentionally and proactively, rather than creating an environment where students feel powerless (*Do I matter? Does anyone see me?*) and subsequently engage in power-grabbing (*Notice me!*) interactions (Boaler, 2008). You can be the leader and authority in the room while sharing the power with students.

Power-sharing is an important aspect of ensuring our classes and schools do *not* become places of punishment and consequence. In the eighth grade (after the incident with Diego I described earlier in this chapter, see page 45), Diego was my classroom aide for one period each day. Our school resource officer and my principal approached me when they were building student schedules and said, "Diego does not get in trouble in your class. We want to give him more time in the day to be successful. Will you let him be an aide?" I wondered, "Is Diego's *behavior* different in my class, or are my *responses to his behavior* different from his other teachers'?" Perhaps it is some combination of both. I do know the traditional system of consequences for students for subjective interpretations of their behavior is a zero-sum game we must stop playing.

Sure, I had plenty of colleagues who shared their advice and opinions about how enforcing school policies *is* teaching students. I was told, "We're teaching them social skills, how to function in group settings, how to play by the rules." I can see that perspective, but I also feel compelled to ask, "Teaching them whose social skills? To play by whose rules?" Rules, routines, structures, and policies are absolutely necessary to help schools be safe places. But when educators inequitably develop and enforce these very same rules, routines, structures, and policies, they can also be what makes schools *not physically, academically, or emotionally safe* for historically overdisciplined students (Howard, 2019).

Reconsidering Disciplinary Action

If students such as Diego are "trees," then looking at statewide summary data is a helpful way to zoom out and get a better view of the entire "forest." Understanding the landscape is essential to informing and improving our decision making about how to respond to students who demonstrate behaviors like Diego. Figure 3.1 is a summary of disciplinary data from the state of Oregon during the 2018–2019 school year (Oregon Department of Education, n.d.). These data include all reported discipline incidents, including infractions, suspensions, and expulsions.

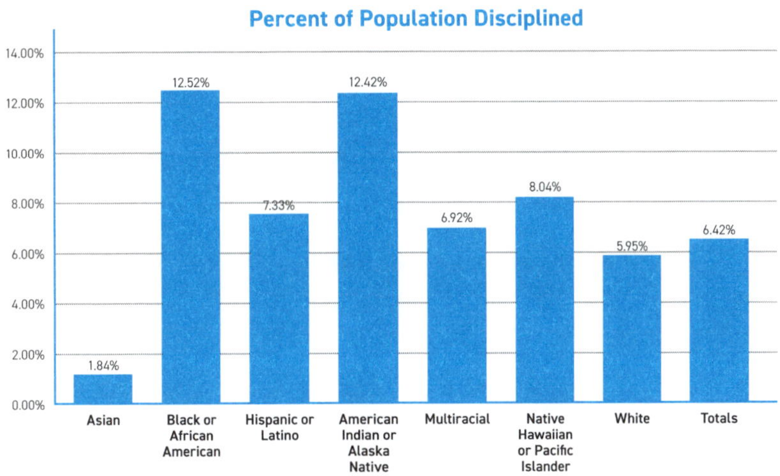

Source: Oregon Department of Education, n.d.

Figure 3.1: Discipline data by subgroup.

PAUSE AND PONDER

When you look at these data, what do you notice? What do you wonder?

You will notice a disproportionate percentage of disciplinary events for Black or African American and American Indian or Alaska Native students—more than double the events of their White counterparts and nearly seven times those of Asian students. Why do you think that is? How do you think things like model minority stereotypes (Lee, 2009; Yu, 2006), implicit bias, systemic racism, and so much more play into the school experiences of your students (Hathaway, 2016)?

Earlier I mentioned *subjective interpretations* of student behavior (see page 54). Take a look at figure 3.2 from the same school year in Oregon. Notice the left-hand bar labeled Disruptive Behavior. This is undeniably a subjective category. How are we defining *disruptive*? According to what standards and agreed-on norms? What warrants a referral or suspension?

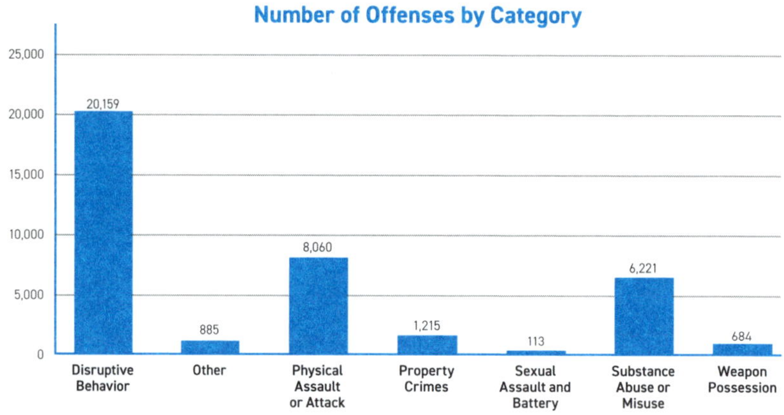

Source: Oregon Department of Education, n.d.

Figure 3.2: Discipline data by incident type.

Since we are in the thick of the data, consider the 2013–2014 data the National Center for Education Statistics (NCES; 2019) shares in figure 3.3. Notice the data summarize out-of-school suspensions.

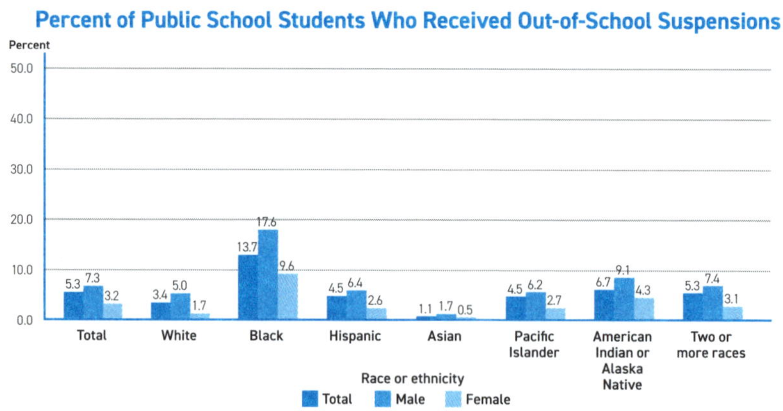

Source: NCES, 2019.

Figure 3.3: Out-of-school suspension data by race, ethnicity, and sex.

The path from single or multiple incidences of disruptive behavior to suspension (in or out of school) is well tread by students still learning how to stop their "dominoes from tumbling." I do not suggest there are no suspendible offenses, but not every disruptive behavior warrants a referral or removal from class. We must check our biases when we are carrying out inequitable discipline policies.

An important reminder about Diego. At the time he was in my class, he was twelve years old. *Twelve*. He was right in the midst of learning how to process and express his feelings, as are your students. We must try to forecast the outcome of our disciplinary decisions, which may do more harm than good. I don't know if the incident in my classroom would have been Diego's first referral or his tenth. I do know the outcome would be time out of class, time out of opportunities to learn, and time to sit and stew all alone, likely feeling there is no place to belong or find any kind of success. What skills would Diego learn while he was suspended or expelled that equip him to make different decisions next time?

Research shows teachers regularly assume Black or African American students are older than they actually are and frequently experience force corresponding to the belief they're bigger than they are (Blakemore, 2018; Cooke & Halberstadt, 2021; Hathaway, 2016; Morris, 2016; Petroni, 2018). Enforcing discipline policies that criminalize students changes their lives forever. Restorative practices that de-escalate and afford opportunities for reconciliation, learning, and growing are instrumental in reshaping traditional school power dynamics (Howard, 2019). As UCLA assistant professor Sean Darling-Hammond (2023), who specializes in expanding belonging by conducting research on K–12 practices that enhance school climates, puts it:

> Educational institutions seeking to implement restorative practices would benefit from encouraging schools and community members to make real shifts in their discipline frameworks (e.g., abating the use of exclusionary discipline) and provide a context that enables staff to shift their teaching philosophies (e.g., relaxing reliance on punitive mechanisms to manage classrooms and encouraging the use of relational approaches). This could be achieved by communicating the negative impacts of exclusionary discipline, modeling relational alternatives for managing common tricky classroom situations, demonstrating the potential of relationships to grow with effort, and providing ample support and time for staff to practice relational approaches before they face classroom conflict. (p. 45)

"You can only expect what you're willing to enforce," one of my professors taught me. I look at current data and reflect on my experience learning how to be an effective teacher for students like Diego, and I wonder,

"If you looked at what *I enforce*, what does it say about what *I expect*? What do my expectations say about what *I believe*?" I encourage you to reflect on these questions about your own practice, and to cross-check your stated values and beliefs against what you enforce.

To Sum It Up

De-escalate zero-sum interactions and restore a learning environment designed for student flourishing and belonging.

As the year unfolded, I learned that Diego and his brother's gang involvement was a top concern for our school resource officer. I learned Diego was regularly detained for in-school suspension and his family life was hard (which he expressed through his drawing in his mathematics journal). I would often swing through the lunch-detention area to say, "Hey," while Diego served out the school-sanctioned consequences for his tardiness, absences, or behavior.

Diego often asked to stay in my class beyond his intervention or aide period. "I like it out here," he would quietly say as he avoided leaving. Typically, I supported him by encouraging him to show up and learn in his other classes. Every so often, though, I wrote him a hall pass, allowing him to be five minutes late, and he would spend that five minutes just outside my portable, sitting on the ramp and swinging his legs over the dry grass. Sometimes he just needed a break; sometimes we all just need a break.

Diego and I created an open dialogue about what made him curious or confused about the content in his other courses. We talked a lot. I listened a lot. He did not need me to "rescue" him. He benefited from opportunities to express his big-life experiences and their accompanying big emotions with an adult who did not require him to soften his tone or turn down the volume of his emotions. I tried to help Diego find tools and connect with our school counselor, who taught Diego invaluable strategies for not getting swept away in the "tidal waves" of his anger. Then, we sent him off to high school, slightly better prepared mathematically and significantly better prepared emotionally.

Students with the most vibrant expressions of their emotions paint a vivid picture of the emotional landscape in which they are growing. As we nurture, protect, and foster their growth, we tend to the conditions necessary for our students (or "trees") to establish deep, anchoring "roots" that will be their sustaining life force for the remainder of their journeys.

Chapter 3 Application Guide

Use the reproducible "Chapter 3 Application Guide" (page 60) to gather your thoughts and make connections between the themes and tools in chapter 3 and your teaching practice. Consider using this application guide collaboratively with colleagues to anchor your exploration, discussions, and application of your learning.

Chapter 3 Application Guide

Chapter Themes and Tools	Connect and Apply to Your Practice
Avoiding zero-sum games	**Learn your hot spots:** Revisit the Learn Your Hot Spots section (page 50) periodically to refresh your practice of recognizing when interactions activate your hot spots. **Gather data:** Create opportunities to connect with colleagues and share formative assessment strategy ideas. Determine how and when to integrate a new strategy. **Create a power-sharing dynamic:** Analyze the current state of power-sharing dynamics between you and your students. Where do you see zero-sum interactions unfolding? How can you disrupt this cycle?

Reconsidering disciplinary action	Gather and analyze discipline data from your school or district to better understand trends that affect different students and student populations. Who would you want to engage in this conversation? How might you set up a process for discussing and determining next steps as a result of your findings? What do you think you enforce? What would your students say you enforce? What does what you enforce indicate about your expectations? What adjustments could you make to calibrate your expectations with your approach to enforcement?

page 2 of 2

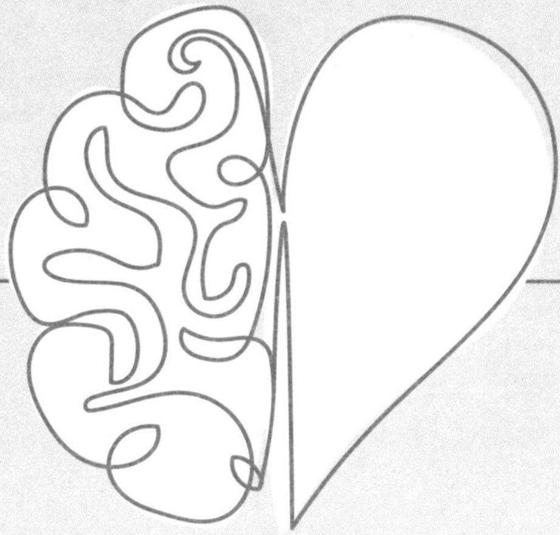

When people feel invisible

long enough,

they will lash out at the world

to remind them that they exist.

—Ralph Ellison

CHAPTER 4

I'm Always in Trouble!

"Ms. Lenhardt, please call the office," the overhead page echoed through the halls as I walked back to my portable after a last-day-of-school assembly. Students were on a modified schedule that day, and teachers were staffing different areas of the school as yearbooks were signed, final farewells were bid, and buses prepared for one last departure. Ducking into a nearby classroom, I picked up the phone and called the office. A moment later, I heard the familiar shouting of a student of mine, "Get the f— away from me!" I immediately knew why the office staff had paged me, so I sat down the phone and headed toward the voice. I had been called to help Amara.

"*Lenhardt!*" Amara shouted as soon as she caught a glimpse of me.

"Heeeey, Amara. What in the world?"

"Get them to *stop*!" she shouted.

All my internal alarm bells were ringing at this point, as I saw our school resource officer, another teacher, and the behavior specialist not far behind her. Without any context of the situation, I was wary, cautious, and concerned. I felt the acute tension between the staff and Amara, and I could see the terror in her eyes.

"Stop what, Amara? What is going on?"

"Mr. Avery kicked me out *again*, so I slammed the door, told him I hate him, *which I do*, and now *everyone is chasing me*."

For what it's worth, I did not see anyone *chasing* Amara. But I do believe she *felt* like she was being chased, and that very real feeling was driving her flight response.

Amara spent a couple semesters with me over the course of her seventh- and eighth-grade years. She had a short fuse and frequently lost her temper. As you've likely experienced with your own students, big emotions and dysregulation often fuel disconnection and disrupt any sort of sense of belonging in many ways. We know big emotions aren't the only facet of any particular student. In Amara's case, she seemed to know a lot about the daily happenings in the student body and from what I could tell, held some social status among her peers. She vocally defended those to whom she was loyal, and had no qualms about letting adults at school know how little she thought of them. I had seen her start fights, and I had seen her break them up. I learned a little more about Amara from her mother during conferences, and it seemed Amara was a force to be reckoned with, both in and out of school.

"Hey," I said firmly and quietly, holding my hand off to the side to signal my colleagues to pause their approach and give her some space. "Listen." She refused to make eye contact, rocking from side to side, a human vessel containing all the pent-up energy of the moment.

"You're safe," I said quietly enough for only her to hear, confidently enough to signal I could help. "Will you walk with me?"

She nodded.

"You good, Lenhardt?" Mr. Ramirez, our behavior specialist, called from about twenty feet away. I flashed him a thumbs-up, and Amara and I started to walk.

As teachers, our interactions with students are often not confined to just our classroom. Across the school campus and in a variety of contexts, we are met with opportunities to engage and connect with students. To invest in cultivating belonging and engagement in our classrooms, we must be prepared to continually respond to the emotional landscape throughout students' broader schooling experiences. Let's explore how the relationships we work to build with students are crucial to our overall efforts to support them in and out of school.

What's Really Going on Here

Amara's story is not uncommon, just as we saw with Diego in chapter 3 (page 45). Many students find themselves eligible for academic intervention as a result of challenges with regulating their emotions. It is a vicious cycle: a student feels academically confused, lashes out, gets kicked out of class for that period or longer, disengages for more instructional time, feels even more confused, and on and on the cycle goes. This was not the first time I had intercepted Amara when she was caught up in the "torrential winds of tropical storm Fury."

Over the course of our interactions, teachers gather information about what helps students re-center and what might exacerbate their anger. We try different strategies, some with more success than others, and do well to continually inquire about the interconnectedness of academic performance and emotional literacy. No doubt, you're thinking of experiences you've had with your own students with a long history of frequent disengagement from learning opportunities, and how it increases the likelihood of poorer academic achievement over time.

Consider the time of year as well, which adds complexity as we combine this vicious cycle of missed academic opportunities and struggles regulating emotions with the rapidly approaching end of the school year. For many students, the approach of summer means the end of school as a steady source of adult support, food, shelter, predictability, and safety. When working to understand what is fueling any emotional "fire," consider that the temporary end of school (whether for a weekend, winter, spring, or summer break) may be rather unsettling for some students.

Whether a vicious academic-emotional cycle, the time of year, or some other complicating factor at the heart of rage, students need help learning strategies to express and regulate their emotions (Siegel & Bryson, 2011). We cannot *always* "calm the storm" students experience. Nevertheless, our role is to keep our cool, model strategies for de-escalation, and provide students with the tools and space they need to calm *their own* "storms." Let's examine the facets of this approach.

Collaborating to Support Students

When trying to support a student like Amara, it's better not to try and go it alone. In my case, the school's behavior specialist, Amara's other teachers, and I frequently met to figure out how we could best support her success in school. The aim of this collaboration is to try to paint the complete picture of what a student is experiencing by sharing strategies with one another, including what has not worked previously. Teachers approach these conversations differently. Some teachers may focus on negative aspects of a student, such as the student's frequent disruptions to learning, skipping class, and so on. Others may share details signaling disengagement, such as the student not doing any classwork, just hanging out in the back (on the physical periphery of learning), or not bothering anyone, but also not engaging in learning. Listen for the ways in which the narrative of a student fosters or hinders engagement and belonging, and for indicators of environmental adjustments we can make to shift in a positive direction.

As you listen, you'll learn which teachers recognize the need to provide empathetic support for your student. Consider banding together and creating a *buddy room* alternative for each other. A *buddy room* is typically a designated alternative location for a teacher to use when removing a student from class. Allow me to explain how buddy rooms functioned in our school.

- **Where to go:** My designated buddy room was my neighbor's portable. I chose this room not just for proximity, but because the neighboring teacher similarly valued fostering connections and relationships with students to re-engage them in learning. I trusted this teacher to support my students when they were there.
- **What to do:** When I used this approach, I would ask the student to take a journal, have a seat at a desk in the back of the buddy room, and write or draw about what the student was feeling. Research shows immune system and heart function improve when people journal to express or recall an event (Siegel & Bryson, 2011).
- **What happens next:** After about five minutes, I would ensure the students in my class were engaged in a learning activity, then I would pop my head into the neighboring classroom (buddy room) and gesture for my student to come back. We would have a brief "let's try again" moment under the overhang between the two portables, and then, together, re-enter class. Taking a buddy room break was temporary, never a permanent "sending away." Students understood the purpose and routine because we discussed it long before anyone ever needed the option. Just like students have routines for academic support, they need routines for emotional support.
- **A note about re-entry:** By the time students reached the buddy room option, they understood what was happening and why.

> Students are smart. They know when they're having a hard time abiding by class norms. This meant I did not spend time explaining the "wrongs" that had led to this option. Instead, I honored the student for knowing exactly what needed to change and simply offered the blank-slate invitation, "Let's try again." This approach conveyed I was not angry, the student was not *bad*, and we could still have a positive experience for the rest of class. It also conveyed we were in this together, sharing responsibility for what happened next.

Be sure to incorporate choice with the buddy room offer. In the face of big anger, consider approaching an individual student, and offering the option of journaling and remaining in class, or journaling in a buddy room.

Buddy rooms won't work for every student. Alternatively, you may need to engage with colleagues to determine options so you can be on call for one another. For example, you may find others with different prep periods; determine who is available to call during each period. This approach is important for students who may need the full attention and support of an adult because it brings that adult to the student, given the student is not in a place to effectively self-manage in a buddy room.

PAUSE AND PONDER

Think about a student who is frequently sent out of class for disruptive behavior. What do you think that student needs help with?

What might it look like to bring help to the student?

I cannot emphasize enough the importance of collaborating with other adults when you are supporting students with a great deal of needs. *Together* is the right way to go. Seek support from your school administrators as well, as they may have unique connections with students, understandings of what worked in the past, or key ways to lighten the load on teachers navigating these circumstances. Due to a few skilled, supportive adults who had wrapped this network of support around Amara and had engaged in frequent discussions about the latest patterns in her behavior, Mr. Ramirez and I exchanged very few words on that last day of school while effectively passing the relay baton from one trusted adult to another. Collaborative foundations established with trusted colleagues equip us to provide together the support our students need. We can help students calm their "storms," re-engage, and continually nurture a sense of belonging in our classrooms.

Creating Context by Telling the Whole Story

As Amara and I walked down an empty stretch of hallway, I gave her emotional space by not asking any questions. In silence, we walked, and after a time, I slowed my pace. Instinctively, she matched my gait, eventually reaching a less-urgent stride. We can use our own pace, breath, and tone to help students co-regulate their emotions.

Her gaze cast down as she walked, she said in a barely audible volume, "He hates me." I crumbled inside. It hurts to be near the pain of others. These moments are so hard. It takes every ounce of self-awareness and strength not to armor up in the face of this pain, which will inevitably push up a student's armor as well. Recognize the secondary trauma stress you may feel by simply knowing about the trauma of another (Hydon, Wong, Langley, Stein, & Kataoka, 2015). When this happens, take a deep breath, stay grounded in your own emotions, and "give kids a chance to tell their stories, which aids them in the meaning-making process that improves their ability to understand their past and present experiences" (Siegel & Bryson, 2011, p. 54).

We continued walking, and I asked Amara the following.

"What makes you say that?"

"I just know he does."

I don't know for sure if Mr. Avery hates Amara. In this instance, I don't think it matters. Our students *sense* they are hated when they get kicked out of classes and feel like they don't belong. This is what requires attention. Be sure to follow up with the other adults involved to get a better understanding of what transpired. In the moment, use your own emotional regulation skills to hold the tension between what may have happened and how it *feels* to your student.

"I'm sorry. I bet that hurts."

"It doesn't *hurt*," she hissed through hot tears. "It pisses me off."

Many of us are not clinical psychologists or licensed therapists. We are, however, human beings, likely familiar with the blurry lines between hurt and anger. For many of us (Amara included), it can be both pain and anger that erupt and leave us needing support. "When the nervous system is *reactive* . . . it's almost impossible to connect in an open and caring way with another person" (Siegel & Bryson, 2011, p. 129).

"Remember the last time you were pissed off?" I inquired.

"No."

"Yesterday. When Alejandra called you that terrible name at lunch. You told me about it when you came to class."

"Oh yeah." (She also used a word to describe Alejandra I won't repeat here.)

"What happened next when you were in class?"

"You let me draw before I joined my group."

"Did it help?"

"Maybe."

"OK," I chuckled. "Should we grab some chalk and go draw out by the portables?"

"You can. I'm not going to draw."

"All right. Will you at least come with me?"

"Fine," she agreed, and we set our sights on Portable 137.

There is a decision point between honoring emotions, staying with them, and then beginning to trek through them. When you're supporting students, attend to the signals that indicate they may be downshifting out of rage. With Amara, a sniffle and perhaps some tears provided evidence of this downshift. Noticing those signals and leaning on our history with students clue us they may be ready to process their emotions. It is important to *not debate* over whether a student's interpretation of events (*Mr. Avery hates me*) is factually accurate. A technical understanding of Mr. Avery's feelings about Amara would not help her reconcile her anger at this juncture. Students' experiences *are* reality, and the moment they're flooded with emotions they don't fully understand or feel control over is not the time to try and correct the record (especially if you don't know what the truth is). Students who feel hated will be angry about it, and that anger is what we work to process.

Notice also the conversation move that redirects students' focus from *this* moment to a previous one, inviting them to remember that while it feels like this moment will last forever, it in fact will not (Siegel & Bryson, 2011). Reminding students of another time they felt the same way, what they did to get through it, and that the feeling did not last forever empowers them to tell the whole story of their experiences. It helps students tap into the many narratives composing their journeys and to not fixate on a partial, reduced version of their realities (Siegel & Bryson, 2011). In the midst of any feeling, students often need help zooming out, lifting their eyes, and re-centering themselves in the whole story of their experiences. It's normal to become hyper-focused on one big experience or one story and have it be the only thing you hear. It takes practice, and sometimes the help of another person, to hear any other narratives.

Providing Support and Outlets for Emotions

When Amara and I arrived at my classroom, I grabbed the chalk for us to use on sidewalks and ramps leading from the main building to the portables. I started to draw, and as promised, Amara did not. When this happens, gauge students' emotional temperature with a few questions.

"Do you want to stay mad?" I asked, authentically curious at her response. This important question helps me know what Amara is ready for. Not what *I* am ready for her to do but what *she* is ready to do.

"Maybe," she sassed.

"So, stay mad?" I asked as I tossed her a piece of chalk.

"*Yes!*"

I kept drawing. Eventually, she crouched to the ground and started using the chalk to make angry tic marks on the sidewalk.

"What are they gonna do, Lenhardt? *Suspend me?* I don't care. I'm not even going to this school next year. They can't do *nothin'!*"

"Yeah, I don't think suspension is the thing." I paused for a moment, then asked, "How are you feeling about the summer?"

She didn't respond, instead focusing her attention on the sidewalk and her chalk.

"Hey," I said gently. "You got this."

She looked up at me and glared. "I know."

"Well, all right then," I laughed lightly. "Now we've cleared that up."

Talking, walking, drawing, and journaling are just a few strategies for processing feelings and emotions (Siegel & Bryson, 2011). Finding a way to exhale is an important aspect of regaining some sense of equilibrium during a "storm." Offering students opportunities to journal or draw for a few minutes gives them something to *do* amid emotions that seem to be controlling their experiences. Students have complete control over what to do with their drawings or journal entries. Some students may keep those drawings or journal entries, but choose not to share. Others might ask to share with you or another adult. Still others may tear them up and throw them away. Knowing they have complete control over the outcome, that no one will "read their diaries," is often enough for students to move past their initial hesitation. As Siegel and Bryson (2011) note, "What kids often need, especially when they experience strong emotions, is to have someone help them use their left brain to make sense of what's going on—to put things in order and to name these big and scary right-brain feelings so they can deal with them effectively" (p. 29).

Several times, I would engage with Amara in writing rather than verbally. It was almost like we passed notes in class. When she needed time to re-center or help getting through a big emotion, I would write a question on a piece of paper and hand it to her. Something like, "Looks like something is up. Need to get it out in the open?" While other students worked through a warm-up or partner task, I would circulate and check in with Amara, reading what she might write back, and responding again in writing. This slowed the pace of the interaction and provided time for me to think about my responses while simultaneously attending to the learning needs of the other students in class.

The power of this strategy lies in both the *medium* (drawing or writing) and the *time* between whatever had happened leading up to that moment and learning I hoped would come next. If we hope to engage students in academics, they need *time* to "let the dust settle." When a student's hot rage is met with, "Leave your conflict at the door," I can assure you it does more harm than good. It sends the message what you require in class is an emotionless experience. It says to students, "You're too much," without supporting the emotional literacy each student is developing. Similarly unhelpful comments include those in figure 4.1.

Directive	Impact
"Go [next door to office or to counselor] until you can calm down."	This directive layers on feelings of isolation and excommunication ("You don't belong here") to an already emotionally tense moment.
"Calm down."	There are no steps or how-tos in this directive. It is stated as though the recipients know how to calm down and are willfully choosing not to.
"Don't slam the door."	Stopping a behavior like door slamming requires an amount of control students need help regaining. The physical expressions of rage will subside when they acknowledge, honor, and process the emotion.
Logic in the face of rage, such as, "I'm sure he didn't mean to make you mad."	Rage-infused debates are futile and create a standoff, motivating students who are feeling angry to defend their position and prove their point.

Figure 4.1: Impacts of some common unhelpful directives.

Instead of comments like those in figure 4.1 (page 71), consider an approach that acknowledges the emotion is real, it is OK to be angry, and there are steps students can take to become *less* angry. Figure 4.2 shows a few go-to prompts for supporting students as they work to calm their own "storms."

Prompt	Results
"Want to write or draw for a little bit?"	Students take their journal, and something to write with, have a seat where they choose, and draw or write. Choose a prompt, write it on a sticky note, and hand it to the student. Journal or drawing prompts include the following. • If you were to describe what you are feeling with colors, how would it look? • What happened? How do you feel about that? • How long do you think you will stay angry? • A student from last period was angry about something. What advice would you give the student to help the student feel OK again? • What has to happen for you to feel better? Then continue with instruction, circling back to this student periodically, re-inviting the student to join the class, and never putting a time stamp on the student's processing.
"Need to get up and move?"	Stand at the door, propping it open, and allow students movement time just outside the classroom. Movement ideas include the following. • Jumping jacks • Push-ups • Running in place • Burpees Or call another teacher or staff member for support, whether it's keeping an eye on the class starting warm-ups, or it's being with the student who needs some movement time.
"Who do you need to support you right now?"	Students often prefer to take a break with the adult they trust most at school, whether the counselor, another teacher, or staff member. You can't always guarantee this chosen adult will be available, but you can try to meet the student's request.

Figure 4.2: Prompts to de-escalate students.

Rather than send the message that learning mathematics requires a disassociation from emotions, how can we ensure students' *whole selves* (emotions and all) belong in our classes? After all, declining to acknowledge or navigate emotions does not mean they are no longer there, nor does it mean they will not play a role in academic learning. Since emotions come to class anyway, let's give them a seat at the table.

I know what some may be thinking: "I don't actually *want* all those emotions from all my students in my class." I get it! It is a lot, but at the same time, it is important. The fear that students will feed off one another's energy and that if we allow one furious student to stay in class, then others will follow suit resonates with me. We may worry one person's tone will ruin the learning for everyone else. What experience shows us to be more accurate, however, is students watch how teachers support their peers and feel an increased sense of stability and security knowing they could expect the same support should they ever need it. The aim is to work to ensure you bring energy to tense situations that neutralizes other (negative) energy and conveys your compassionate, empathetic care for each student. A single student could not rock Portable 137 off its axis. Students take comfort in knowing the classroom and its adult representative will work through whatever "storm" rages.

PAUSE AND PONDER

How do you provide students the buffers they need (in time and activities) to start to process or regulate their anger?

If you're worried some students might take advantage of such an approach, you can keep track of those data too. Are students taking advantage of your compassion? Are there hints of manipulation? On the weekly printout of each class roster, consider marking an asterisk next to students who need any portion of class time to cool off or take a break. Jot down an estimated number of minutes they need before they're able to rejoin typical class activities. Periodically review your data. The hope is to see student-specific data decrease over time as you collaborate to develop their skills to re-center when their emotions are big.

In the event you notice data trending the wrong direction, consult with specialized staff members, including behavior specialists, school counselors, and other teachers with their own unique ways of interacting with students. In principle, it is essential we hold tightly to the idea students *can* learn how to regulate their emotions and very loosely to the approaches that might bring this to fruition. Continuously gather strategies and ideas from others, and then combine them with your own intuition and innovations to try to meet evolving student needs.

I remember fewer than five students over the course of my in-classroom teaching journey who were, in fact, taking advantage of opportunities to journal or draw. It came to light within about a week of steady opportunities and data gathering. In each instance, I talked with each student outside class to learn more about what was going on. I found the following conversation starter particularly useful.

"I want to be sure I'm helping you get what you need when you're in my class. You can ask for what you need. I don't think when you're asking for a break, it is really helping you, and I wonder what that's about."

Ample conversation space following this prompt usually led to honest conversations about the academics, social dynamics, and all sorts of things not on my radar. Students each shared that the content we were learning was too easy or too difficult. They shared about social dynamics with exes or former best friends that were impacting our class time. The point is, by noticing what was not working and providing students opportunities to fill me in, we found solutions, students felt seen and heard, and we were ultimately able to re-engage in learning. Figuring out how to encounter, attend to, and support the emotional needs of individual students requires an ongoing commitment to learning skills outside the field of mathematics.

To Sum It Up

Embrace the de-escalating power of empathy; model and teach emotional literacy.

"What if I can't do the math in high school?" Amara asks seemingly out of the blue. Except it isn't really out of the blue if you lean on that connected understanding of her anxiety approaching the summer, paired with cyclical behavior and academic challenges.

"You tell me. What are you gonna do?" I volley the question back to her.

"Get kicked out of class so I don't have to do it," she grins. She and I have talked about this many times before, and she fully understands how to use the structure and policies of school to get out of something uncomfortable or hard.

"That's one approach," I say, not taking the bait.

"Or I could get help," she says rolling her eyes in her unique *yeah, yeah, yeah* fashion.

"Yep, or you could choose that."

"What if all my teachers suck?"

"Do all your teachers here suck?"

"No."

"I bet it's the same in high school. You may not like them all, but I don't think they will all suck."

"You get me, Lenhardt."

"I sure try, Amara. I sure do try," I gently respond, affirming she is, and will always be, worth the effort to figure out.

She stands up, drops her chalk into the basket, and sighs.

"Ready?" I ask, as I do the same.

"I suppose," she replies. We trek back to the bus lines together.

I see Mr. Ramirez headed in our direction and nonverbally signal to him we are all right. As he walks by Amara, he puts his hand up, and she high-fives him in an unceremonious departure from middle school. Amara's whole story is much more than her "tidal wave" of emotions that day. As a team of caring adults, we reminded one another of that, and committed to reminding Amara too. She is *more* than the sum of a series of volatile moments. She is *more* than the academic challenges she worked to overcome. We all are more than one track, more than one story. Embracing the multiple, intertwined threads of our own stories keeps us in touch with the multidimensional nature of others.

Chapter 4 Application Guide

Use the reproducible "Chapter 4 Application Guide" (page 76) to gather your thoughts and make connections between the themes and tools in chapter 4 and your teaching practice. Consider using this application guide collaboratively with colleagues to anchor your exploration, discussions, and application of your learning.

Chapter 4 Application Guide

Chapter Themes and Tools	Connect and Apply to Your Practice
Collaborating to support students	Who on your team of caring adults is committed to empathetically supporting students learning how to regulate their emotions? Note the staff (including those not on your team) with whom you can partner to support students who are not yet thriving. Is there an existing process to support students? If not, start small with a like-minded colleague you can partner with to brainstorm about how best to support students.
Creating context by telling the whole story	Put a note to yourself in your planner or in the back of the room where you will see it (for example, "Tell the whole story" or "Add important context"). This note is to remind you to create context about your students' current situation to help you empathize and help students through their most-intense emotions.

Providing support and outlets for emotions	In what ways do your students demand you see them? What does an "emotion storm" look like? How might you craft opportunities for students who need it to write, draw, or move? What would those opportunities entail? How will you gather data about the effectiveness of your approach?

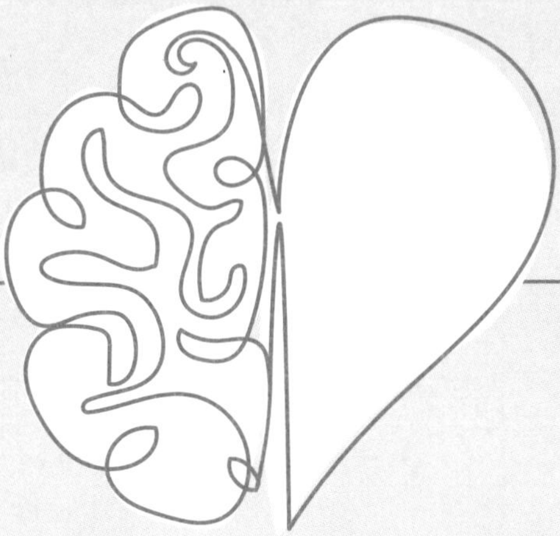

Teachers can inspire students
beyond or apart
from difficult life circumstances,
and they can take advantage of strength
that all children bring to school.
They also can disrupt or eliminate,
rather than perpetuate negative images . . .
and beliefs about who can
learn mathematics.

—Julia Aguirre, Karen Mayfield-Ingram, and Danny Bernard Martin

CHAPTER 5

I'm Not a Math Person

As was often the case during class, I was mingling among students while they worked, listening to their discussions about mathematics and their experiences learning it. A few students were scattered at tables more to the periphery of the classroom, and I mentioned to them individually that in a few minutes, I would have them join a group to discuss.

I have learned over the years that *where* students choose to sit is a piece of useful formative assessment information. Seating choices sometimes offer clues about readiness, perhaps even willingness, to engage in learning that day. Are they sitting up front? Off to the side? Nearest to the exit? With others or alone? You get the idea. Where a student chooses to sit doesn't always mean something, but it might, so it's important to notice.

As promised, there came a moment in the lesson when it was important for all students to engage with one another to process their learning. I made my way to a table in the very back where only one student was seated. After asking her to join a group, she huffed, smirked, crossed her arms, maintained her gaze at her paper, and firmly said, "I'm kind of a loner."

Oof. A tender wave of knowing and shared experience swelled at the center of my being as I thought, *You are so much more than a loner.* I smiled in return and said, "If it's OK with you, I'll circle back when groups are shifting again, and if you'd like to talk with someone, we can make it happen."

"I guess. Sure." She looked up, and her eyes met mine.

As I stepped away from her table, I wondered about a more global implication of this exchange. There are moments when students say aloud the messages they are internalizing about who they are or who they are becoming—moments where we catch a glimpse of how those internalized messages are beginning to solidify identities and drive what students do. Regaining my attention to the present moment, I feel afresh the intensity of the "who I am–what I do–who I am" cycle present in each student's ongoing development of multiple identities. What might I do to ensure the routines, structures, and interpersonal dynamics in my classroom cultivate the types of positive, identity-building experiences students need?

What's Really Going on Here

How many times do we hear from students, strangers, and even ourselves statements of *behavior* spoken as statements of *identity*? In the case of this student, her self-described status as a "loner" seemed to be the loudest story she told herself in that moment. Perhaps this is because I was asking her to potentially move to interact with others. Did she feel like she belonged there? How else might she have phrased this? "I'm kind of a loner" versus "I prefer to work alone"?

Words are important. They matter *a lot.* The words we choose, even when we are being careless, reveal our beliefs and help us gain insight into the beliefs of others. Words are the manifestation of our beliefs about the world, and more pointedly, what we believe about ourselves.

Whether we are in the role of the teacher or the student, in school or out in the world in another capacity, we hear the lines between behavior and identity muffled with statements like the following.

- "I'm such an idiot" versus "I made a mistake"
- "I'm not a good student" versus "School is hard for me"
- "I'm not a math person" versus "I don't like math" or "I have to work hard to understand math"

The last item on the preceding list, the infamous *I'm not a math person,* is one you have certainly heard as often as I have, and a sentiment we will focus on in this chapter. Each time I hear it, I feel compelled to politely disrupt the assumption that being a "math person" makes you a different sort of being. What does that even mean, to be a *math person*? Does it mean you like mathematics? Or you are curious about it? Or you enjoyed learning it? Or you are fast with facts? I honestly cannot tell you, and this is my chosen field of study! "What does it mean, to you, to be a math person?" I usually reply with genuine curiosity.

The student responses vary, but most share an undertone worthy of attention. Typically, replies about being a math person include something about getting along with numbers or being fast at calculations. However, when it comes to *not* being a math person, common refrains include, "I was fine in algebra, but once I hit geometry, it was over for me," or something along those lines. It is all too common for students to share their mathematics learning journeys were truncated prematurely or express their calcified beliefs they aren't able to learn or understand mathematics the way they may learn or understand other subjects (Aguirre et al., 2013; Boaler, Wiliam, & Zevenbergen, 2000; Kuo, 2016).

PAUSE AND PONDER

What emotions surface for you when you think about the statement, *I'm not a math person*?

What about when you hear, *Learning math is hard for me*?

Each time these discussions occur, I experience a proverbial gut punch. We want so badly to wave a wand and remove whatever experiences led the student in front of us to develop a belief that the world is filled with mathematics and non-mathematics people. It is never too late to speak to this false binary, to gently and intentionally chip away at those hardened beliefs. Whenever you can, lean in, respond, and try to equalize the intellectual playing field that seemingly tips in favor of those perceived to have mathematical abilities.

Consider offering some version of, "Learning math is hard. It is hard for everyone at some point. It was hard for me too."

You may be surprised that with that response, you normalize struggle as a part of learning and may hear something like, "No one has ever said that to me before," in reply.

We already know how true this is. If the student speaking to us had already encountered this perspective, then the student likely might have a more growth-oriented narrative such as, "Learning math is hard for me," instead of, "I'm not a math person." What does all this have to do with belonging and engagement? It is the observable pattern; when we are worn out from our mistakes, feel defeated by the challenges before us, or overwhelmed in our efforts to overcome obstacles, we resign to the circumstances surrounding our struggle and adopt a fragile mathematical identity. Notice among your students, perhaps among your peers (or even within yourself), how when encountering challenges, people sometimes internalize emotions in a way that starts to sound like, "I don't belong in math" or "I won't make it in this STEM career" (Kuo, 2016).

Think about the students seeking belonging while forming their identities and learning about their gifts, as well as developing their skills. How early in this process do so many students decide that because learning mathematics is difficult, they must not be "math people"? As teachers, we are in a position to disrupt these developing storylines by speaking to a more wholistic truth. How might this look? How do we continue to help students see who they are becoming while honoring the complexities of learning something that does not come easily without inadvertently reinforcing the self-limiting narratives we hear? Some of the answers we seek lie in reimagining the cadence and structure of learning to better meet the needs of students throughout their time in school, given the likely increased anxiety they experience when engaged in mathematics in social settings.

Let's take our cues for how to engage in this work from what we know about long-distance events like ultramarathons. Although I have not run a solo marathon, I have prepared for and participated in a locally notorious 130-mile walking relay race from Portland, Oregon, to the Pacific coast. Allow me to describe some of the relay dynamics that are particularly helpful in shaping a sustainable approach for our students learning mathematics.

Drivers and Walkers

At the start of a walking relay race, one walker takes off, and then the walker's teammates pile into a big van and drive to the official exchange point, where the next walker starts warming up in anticipation of the relay handoff. The key logistic point is each team consists of several walkers and one driver. The driver's job is to transport walkers, follow the maps and signage, and arrive in a safe and timely manner at various checkpoints along the way. The driver also takes on the role of *chief morale improver* (that is a technical term, thank you for asking). The driver's demeanor makes a *significant* difference in setting the tone for the team. Whether by choosing music to energize the team, or driving in silence while others sleep, the driver regularly attends to the needs of the group. On the long drives stuck in a line of traffic or figuring out a location when there was no cell phone service, our driver cracked jokes and told stories to keep us distracted from the enormity of the distance yet to go. As a walker, I came to trust the driver's expertise and knew I could count on him as a crucial member of our relay team.

You may already see the parallel here to teaching. In this analogy, the teacher is the "driver," the one responsible for knowing the route or having maps ready, for ensuring the timely and safe transport of the students ("walkers"), and for maintaining or improving morale in the hardest stretches of the learning relay. Articulating the difference between the role of a driver and the role of the walker helped me better understand my role in supporting students. However, it is my experience as a walker that enables me to empathize with students who seem ready to surrender to *not being a math person.*

I recall what it feels like to be completely exhausted and uncertain of my ability to finish what I set out to do. I recall sore, tired, and cramped muscles on the third leg of the race. I remember the blisters that developed from tiny rocks lodged in my shoes. The actual event participants (walkers) put forth tremendous effort, sustained injuries, and often limped rather than sprinted toward the end of the race. A few additional race-day details are useful when considering how to support walkers through grueling times, and they all have applications to supporting students learning mathematics: water stations, encouragement, nourishment, interval training, and needs assessment.

Water Stations

Long-distance events such as marathons and relays follow a predetermined course. For any major event, you can expect to find medical aid stations, water stations, and places to rest along the route. As an event participant, you come to rely on those rest points as mini-finish lines, places that mark your progress and provide essential sustenance. I remember the feeling of seeing our van parked along the route or seeing the water station up ahead and experiencing a flood of "I'm getting there!" energy. Each ensuing stop marked progress in the right direction.

Exhausted from class after class, year after year of struggling to learn mathematics, many students need a "water station" or a break to refuel along their "ultramarathon" of learning. Within a stretch of learning, whether a class period, a semester, or school year, where do you provide water stations for your students? What do they look like? The following details the places I learned to strategically offer students water stations throughout the long stretches of learning.

WATER STATION ONE: WITHIN A CLASS PERIOD

Encourage students to "stop for water" when you transition from one station or activity to the next. This pause may include an extra minute to talk to someone in a different group, a check-in with the teacher, standing at the door or window for a moment to see what is happening outside, and so on.

- **Purpose:** The purpose of this activity is to provide students frequent, but brief breaks—opportunities to process their learning as well as to physically move.
- **A concern:** Transitioning takes too much time, or it will be too difficult for students to refocus on learning after you have allowed them time to pause.
- **And a thought:** I found when students knew the routine, cadence, and expectation of *learn a little, rest a little*, we were successful in re-engaging and not getting off task during transitions. I only experienced lost instructional time when I had not communicated those expectations well at the outset. I even used the image of water stations when I talked with students about what a class period of learning would look like, explaining that the breaks were pauses along the route, not the finish line.

PAUSE AND PONDER

How might you provide a "water station" within a lesson? What would it be? When would you provide it?

WATER STATION TWO: ACROSS A UNIT OF INSTRUCTION

Insert *mathematics buffers* between days of particularly challenging content. A mathematics buffer is anything you can insert into a lesson with a heavy cognitive load to provide a break in the hard work of learning something new and provide students regular opportunities to find success and fun. These buffers may relate to other mathematics principles students feel more confident about, or even non-mathematical content. In my experience, students benefit from this type of buffer about every five class periods or so, although sometimes more frequently.

Be prepared to use a mathematics buffer on the fly (for instance, when a quickly derailing lesson leaves you not entirely sure how to proceed to get learning where it needs to be). When this happens, leverage the opportunity for some mathematics buffer time, providing students (and you) the necessary opportunity to rest and replenish their mathematical (or teaching) muscles. I share suggestions for mathematics-related buffers in chapter 7 (page 117).

A second type of mathematics buffer provides an opportunity to explore student questions that are not necessarily mathematics related. In service of reigniting student curiosity, create a space to surface and reinforce students' natural curiosity. I called this space in our class the *wonder wall*. Students could write whatever they were wondering about on note cards and post them on the wall. During a mathematics buffer, we collectively chose something off the wonder wall and had a discussion. Students' wonderments included a full range of topics such as, "I wonder what is in the school lunches," and "I wonder what I'll look like when I'm older."

- **Purpose:** The purpose of a wonder wall is twofold. First, it promotes curiosity and encourages students to bring their entire selves and experiences to mathematics class. Second, it helps students learn there are no off-limits wonderments.
- **A concern:** What if students ask something inappropriate?
- **And a thought:** Try not to be shocked when students post something inappropriate. Use your discretion, and in some cases, just take the card down without discussion. Exceptions are hurtful, targeted statements like, "I wonder why Miguel is so ugly." Imagine, if this was how poorly Miguel was treated with an adult in the room, then what was happening when no adults were around? Responding to bullying statements or other equally toxic statements is a crucial aspect of cultivating belonging and ensuring social safety. How to approach these conversations is situational and complex. The important thing is not to be silent or dismissive of events like this. Hurting other people is not OK, and when it occurs, it warrants

discussion and processing with the whole class, leveraging your school's resources for anti-bullying lessons, teaching kindness and respect, and so on.

PAUSE AND PONDER

What might a mathematics buffer look like in your class? How can you tell when you or your students need one?

WATER STATION THREE: ACROSS A SCHOOL YEAR

Incorporate a check-in with students and see how they're doing whenever you return from a school break or when emotions may be running hot. Depending on the year and your particular students, you could check in on Monday mornings after summer, winter, or spring breaks; after conferences; after an assembly; after a lockdown; or after an upsetting community event.

Check-ins may consist of whole-class or small-group discussions or independent opportunities to write or draw. Remember the journaling opportunities that were so helpful to Amara in chapter 4 (page 63)? Periodically, those same opportunities make sense for the whole class to help students navigate complex emotions that arise due to various circumstances.

- **Purpose:** The purpose of this activity is to listen to students, which is a trust-building move (Hammond, 2015). Building trust with students is an investment in learning partnerships, an essential part of cultural responsiveness that helps us attend to students' learning processes (Hammond, 2015). Essentially, whether through discussion or journaling, this activity provides a bridge between students' current emotional contexts and academic learning.
- **A concern:** "There is always something going on in my school or in the community. There is no way I can pause for a 'water break' every time something happens."
- **And a thought:** Remember, the "distance event" stretching ahead of students is for *the rest of their lives*. As teachers, we are modeling how they can pause along the way and take care of themselves in preparation for the "next leg." The way I see it, while stopping frequently may make it seem like we will never

get to the "finish line," be assured if we *do not* stop for breaks, we also will never make it.

PAUSE AND PONDER

When during the year do your students need a "water station"? Do you ever try to push through without stopping? Why?

Encouragement

Along the route of the relay race from Portland to the coast, I experienced encouragement from my teammates as well as from other walkers. I also had the opportunity to be an encourager. This was a competitive race, yet when I passed another walker, I would encourage with a shout, "You're doing great!" or "Almost there!" I received the same sort of support when another walker would pass me. Encouragement is a learned behavior for most of us. We teach our students how to be encouragers by modeling how to do it. The following are a few encouragements that speak to students' hard work and remind them their efforts will pay off.

- "Look at how much you have learned!"
- "I can see your dedicated work ethic in full force today!"
- "Yesterday, this concept was still confusing to you, and today you understand it better. Nice perseverance!"
- "So smart of you to take a break. It's always helpful to notice when you need to pause."
- "I know you can figure this out. You are not alone; your whole team is here to help."

Remember to counter narratives like, "I'll never learn this," with statements that honor the complexity of the learning process and normalize feeling weary along the way. Saying something like, "Sounds like you're feeling pretty tired from working on this. I know you can do it. How about a break?" helps students know everyone gets tired from time to time, especially when trying to learn something new or challenging.

Perhaps an ounce of caution is in order when we talk about how to encourage students, a conversation that often includes mention of a *growth mindset*—the relationship between hard work and dedication and our developing abilities (Dweck, 2008)—and *grit*—the outpouring of passion and perseverance (Duckworth, 2016). To be clear, *supporting students in developing a growth mindset* means conveying to them the brain *can* learn new things. It is not about positive thinking our way out of systemically challenging or oppressive situations (Hammond, 2015). Similarly, when it comes to *grit*, be mindful not to solely place the responsibility for sticking with it on the student experiencing the challenge without properly analyzing the structure of experiences that may make it impossible to keep going (Sultan, 2015).

PAUSE AND PONDER

What obstacles are teachers (as well as the other adults in school) responsible for removing from students' way?

How would you describe the impact of those obstacles on the students' learning exhaustion?

Nourishment

The van for our Portland relay race team was packed with huge coolers of water, sports drinks, and foods to keep us hydrated and nourished throughout the race. One of my post-leg absolute favorites was an egg salad sandwich. (We don't have to agree here, but I can assure you they tasted so good after a long, physically and mentally exhausting several-miles relay leg.) An egg salad sandwich is a prime example of the less-than-perfect nutrition that contributed to a successful race. In addition to the foods with high nutritional value, we also planned for, packed, and consumed foods only marginally important nutritionally, but significantly important for morale.

It is important to provide this sort of nourishment in the classroom as well. Not everything we learn or do in class is a "protein shake," and not everything is "candy" or "chips." Within each lesson (or across a stretch of time you have with students), give yourself permission to include something fun and lighthearted, an academic "egg salad sandwich." In my classroom experience, I found these lessons or learning moments were often what students looked forward to most. They were the "soul foods" keeping our spirits high and a reward waiting at the end of an exhausting "leg of the race." *What* we are learning matters as much as our *experience* learning it. A few of my mathematics-intervention egg salad sandwiches included the following.

- **Writing warm-up:** Provide students with a list of quotes and ask them to write about one that stands out to them. Try to select quotes that represent a variety of sentiments and include people students are familiar with, such as music, sports, or political legends, or contemporary social media influencers.
- **Name patterns:** This mathematics activity focuses on finding patterns in students' own names. Students each use a 6-inch by 6-inch grid to write their first name repeatedly, then highlight the first letter in the grid. Then, they compare their own patterns with those of their classmates. In my classes, many rich mathematics discussions resulted from this initial exploration. And in their semester reflections, students identified this as one of their favorite, most memorable learning moments.
- **Pattern block art:** Before introducing pattern blocks in a lesson, provide about fifteen minutes for students to explore the pattern block pieces and make artwork with the colorful cutout shapes. Students enjoy the creativity, and their artwork can decorate the class walls for a few weeks.

Aim the learning experiences you design as nourishment for *how* students learn. Remember what it feels like to be in class, and find something for your students to look forward to. This is an important detail to include when creating a space where students each belong and helping them make it through the "ultramarathon" stretches of learning.

Interval Training

If you plan to make it through a long relay or distance event of any kind, it is wise to prepare. Training in smaller intervals is key to building the physical capacity you will need. In addition to the physical benefits of interval training, there are mental benefits. Training in intervals using low-stakes, shorter sprints is how you experience intermediate successes that fuel future efforts. The memories of these mini-wins will serve you when you're in the final half-mile of a grueling leg and uncertain if you can make it. *You've made it a half-mile before, during practice and training; you can do it again!*

The same is true for building the strength and endurance to make it through long stretches of academic performance (think standardized, high-stakes tests) or for the more ultimate distance students each will traverse over the course of their life. Knowing you can succeed at three minutes will help you believe you can succeed at thirty, or during your 6.8-mile leg of a race. Success breeds success. Shortening the time frame of activities and working for smaller wins are ways we can begin to rewire students' internal stories of failure. Challenges they face often fall into two categories: (1) academic and (2) social-emotional. Both types of challenges require interval training. In the following sections, read about two of my go-to practices—academic intervals and social-emotional intervals—and make notes of your own ideas too.

ACADEMIC INTERVALS

Start with short tasks. Consider posting one problem for discussion on the board, rather than having students turn to a page filled with the outline of the entire day of learning. It can be overwhelming to know there are "five miles to go." Assure students that, for now, we are just going "down the street," and then we will go from there.

Consider the following examples as you determine how to structure your academic intervals.

- Transfer a set of problems to individual note cards. Ask students to work in groups. Each group gets one note card with a problem to solve. After solving the problem, a student from the group gets up and checks the group's answer with another group. If disagreements about process or answers arise, the whole class may engage in figuring out the solution, with your support and guidance. This process is important; it helps surface any partially developed understandings requiring your attention. The student then retrieves another note card, returns to the group, and together the group members tackle the next problem.
- Create a gameboard of tasks or problems (think bingo or in the style of *Jeopardy!*). Students can work in small groups or as a whole class to try to "win" the game by selecting the next problem to solve.

In both examples, your academic-interval structure changes the feel of typical practice with problems, removes the pressure to get through all the problems, and provides space for students to focus on the immediate stretch of learning. It also allows for intermediate celebrations, which inherently boost engagement.

PAUSE AND PONDER

How can you break up academic tasks into small intervals?

SOCIAL-EMOTIONAL INTERVALS

Mathematically rich discussions are essential to developing deep understanding (Hattie et al., 2017). To develop the stamina students will need for sustained, focused conversations, start small with non-mathematical topics. The following are a couple of examples, along with an opportunity to brainstorm one of your own.

- Turn to a partner and take two minutes to talk about the next movie you want to see.
- At your table group, take three minutes to discuss and find one thing you all have in common.

PAUSE AND PONDER

How can you engage students in non-mathematical stamina-building opportunities that will serve them mathematically in the long run?

Needs Assessment

Prior to race day, our Portland relay team organizer asked each of us, "What do you want from the van when it meets you along your leg of the race?" It was a straightforward invitation to consider what would be most useful for us, to communicate it to the other people on our team, and to know on race day our needs would be met. Students, when offered the opportunity, are also capable of learning to describe what they need. This process includes modeling how to ask for what you need and holding space for students to do the same.

- **Modeling:** This may sound like, "Hey, team. I am having a hard time focusing today. Thanks for your patience if I seem a little scattered. I'm going to take the first three minutes while you do your writing warm-up to get my head in the game for our class."

- **Holding space:** Not everything has a formula, but this structure is useful—Name what you notice + Invite a student to name the need. "Seems like you may be running on empty. What do you need to refill your tank?" (Think water station, encouragement, and so on.)

To Sum It Up

Invest in identity-building and structured learning for endurance throughout the learning process.

Back in class (with my student who chose to sit at the back of the room), as promised, I circled back and invited her to join a new group to discuss the next part of the lesson, which she did. Message the following in your repeated invitations.

It matters what you have to say. I can drive you to the next spot in this relay, and you have teammates who are there to cheer you on.

And keep circling back and extending invitations no matter how many times your students decline. No previous response changes how worthy a student is of the next invitation. You will likely begin to notice small intervals of success, including times when students, of their own choosing, do connect with their peers and engage in learning without your prompting. It may be brief, but it is a start, and these intervals build momentum.

At the end of class, as I stood at the door and bid farewell to students, the student who sat at the back of the classroom lingered at the back of the group.

"Thanks for not giving up on me," she whispered as she walked out of class.

"You betcha," I whispered back.

I have exactly zero knowledge about the entirety of this student's mathematics story, or of *her whole story*. I do know what it sounds like when students are questioning their belonging and unsure of their stamina to stay engaged throughout a learning challenge. It can sound like, "I'm kind of a loner," or "I'm not a math person." It also sounds like "the race getting the better of the contestant," and this often requires external support to write a different ending to that story. When you are creating environments where students know they belong and have what they need to get and stay engaged, you should include "water stations," "soul food," and others "cheering them on." And, your students will periodically need you to be the "driver."

Chapter 5 Application Guide

Use the reproducible "Chapter 5 Application Guide" (page 94) to gather your thoughts and make connections between the themes and tools in chapter 5 and your teaching practice. Consider using this application guide collaboratively with colleagues to anchor your exploration, discussions, and application of your learning.

Chapter 5 Application Guide

Chapter Themes and Tools	Connect and Apply to Your Practice
Using water stations	Revisit your notes from the water station examples (see page 83). Ensure students have water stations (or refueling opportunities) within a lesson, within a week, and across a unit of instruction. Connect with colleagues for ideas, or consider engaging students in a discussion to gather their input.
Differentiating your encouragement	Get to know how your students prefer receiving encouragement. Create a survey or inquire during one-to-one check-ins.
Providing nourishment	In addition to *what* you plan to teach, consider *how* students will engage in learning, and ask them to share how they *felt* during the learning experiences. Incorporate their feedback into an upcoming lesson (or set of lessons), explicitly sharing that you designed the lesson with their feedback in mind. Always remember to consult with other teachers to generate new ideas as needed.
Using interval training	How can you break up academic tasks into small intervals? How can you engage students in non-mathematical stamina-building opportunities that will serve them mathematically in the long run?

page 1 of 2

Modeling for students how to ask for what you need	To what extent do you regularly articulate and model for students what you need? How can you integrate additional modeling into your practice (if needed)? Prepare for upcoming interactions with students by jotting a note or posting a sign where you can see it that reminds you to "hold space."

PART 2

What We Say to One Another

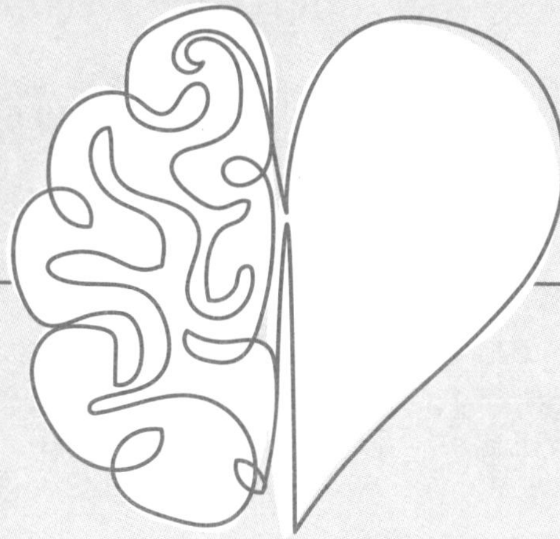

I am all out of Fs.

—Jennifer A. Lenhardt

CHAPTER 6

I Don't Give Grades; My Students Earn Them

Let's talk about grading. At the start of my third year of teaching mathematics intervention, our school was under a new administration. The new principal wanted to increase each teacher's accountability for the *F*s issued on report cards. His general initiative sounded something like, "If we are assigning students *F*s, we should examine our teaching and take responsibility for how *we* are failing our students. Do not view it simply as students failing our classes."

On the surface, I can get behind the message: we do not relegate our students to cycles of failing by stamping *F*s on their report cards. I will dig into the nuances here in a moment, but first consider the opposing sentiment—something that often sounds like, "I do not *give* grades; my students *earn* them." This, admittedly, is an outlook I *cannot* endorse.

A more careful examination reveals the similarities between these two sentiments. One places complete responsibility on the *students*, and the other places complete responsibility on the *teachers*. Neither addresses the problematic, ineffective system of traditional grading, which leads to teacher and student tracking (NCTM, 2018). *Student tracking* refers to the course pathways or "fixed sequences" that provide different levels of access to mathematics learning (NCTM, 2018, p. 16). *Teacher tracking* refers to the common practice of assigning the least-experienced teachers to entry-level mathematics courses and "those perceived to be the most effective, assigned to upper-level mathematics courses" (NCTM, 2018, p. 21).

For the success and thriving of teaching and learning, it is critical that we develop equitable, rather than punitive, systems for measuring student learning. When I revisit these ideas, I see the ways my school's policy fell short in terms of equity. Our grading practices and systems at the time were not set up to reflect students' individual and personal growth academically, socially, and emotionally. Many changes were needed, and although I was unable to redesign the approach to grading for the entire district, I did stand up for my students regarding what I *could* control. Allow me to share the following story to illustrate what I mean by *inequitable grading systems.*

I had an interesting encounter with this new principal when it came time to turn in semester grades. I had submitted my grades by the designated deadline. There were probably six or seven students for whom I had entered a grade of incomplete (*INC*) rather than submitting a letter grade. The principal approached me during our grading day and shared that while teachers could submit INCs at the six-week grading interval, our grading system would not permit INC grades for the semester. I asked, "Then what do I change it to?"

"Well, what would their letter grade be?" he asked.

"Nothing. I mean, the students we are talking about have not had a chance to demonstrate learning or growth."

"What do you mean?"

"Several have been absent much of the semester, one due to out-of-school suspension, one due to illness, and the others, I am not sure of the reasons. Also, many of the students I marked INC have not had a chance to revisit their assessments, and so on. Their work is incomplete, which is why I marked it as such."

"You will have to change them to *F*s then, in the system. Otherwise, your grades are not considered finished."

"But they did not earn *F*s. I don't give *F*s."

"But they have not met the requirements to earn anything other than *F*s."

"Yeah, I see what you're saying. But I'm not doing that. I am not going to say they have failed. It doesn't reflect the situation and doesn't align with what I'm trying to achieve with intervention. Why can't the grading system allow an exception?"

"That is just how the program works—it's a computer thing."

Are you experiencing the raised heart rate and tension I felt during this exchange? As my brow furrowed and my arms crossed, my protective instincts started to simmer. I felt responsible for my part in the system that would potentially harm students on the receiving end of report cards that misrepresented their learning journeys. I could see the faces of the young students who would receive their report cards and see, in an *F*, confirmation they were *failing*. The weight of that grade, an *F*, on the growth and development of young students was unacceptable, and I made my disagreement known.

"Well, I'm not changing the marks. I am not giving them *F*s for work that isn't complete, for students who have not had the opportunity to learn."

"So, you're just not going to follow the policy for submitting grades?"

"I am not. It is not the right policy."

"You will probably be written up."

"OK."

And so, I was. It was not a serious disciplinary action, but it was noted as some measure of insubordination in my file. And in future conversations at this school and with other teachers, we started to move toward standards-based grading and advocating with the district for technology within the grading system that accounted for incomplete grades for students whose learning was just that—*incomplete*. Whether it is a grading system, making mathematics enrollment recommendations for students from year to year, or disciplinary action, we have an opportunity—indeed, an obligation—to examine practices and policies. Investigating, questioning, and advocating for students to have ongoing opportunities to engage in meaningful mathematics learning opportunities are moves we can make to cultivate engagement and belonging. Let's use this example of report cards and *F*s to launch our exploration.

What's Really Going on Here

We must consider students' futures when we make decisions about their present school experience. When we give *F*s, there are often serious implications for students that can alter their learning trajectory. Students' report cards and grades are often criteria districts use to determine their class placement the next year. In a system with gifted and talented, regular, and lower-level mathematics course pathways (based on tracking), students with the lowest grades are typically placed in the lowest track (NCTM, 2018). Because learning in different tracks is qualitatively different, the learning outcomes are inequitable, having "long-term negative effects on achievement and affective outcomes for the students in tracks with less access to rigorous and high-quality instruction" (NCTM, 2018, p. 18).

We must trust what the data show about grades and student tracking. We know learning opportunities for students in different ability groups (low, middle, and high) are substantively different (NCTM, 2018). We must trust what the data tell us about *teacher* tracking as well and recognize the district assigns the highest tracks of classes to the teachers it judges to be the most competent and experienced, or who carry the highest status. By contrast, the district assigns classes with the lowest ability track to teachers with the least experience (NCTM, 2018).

We must make decisions *now*, such as refusing to give *F*s for grades, to disrupt that self-fulfilling path to academic failure. We must advocate for our students by working with colleagues to revisit established grading practices and critically evaluate their effectiveness. Although the progress toward systematic and systemic change may seem slow at times, engaging in this analysis and shifting our approach is important work.

So, what is interfering with the establishment of equitable grading systems that reflect our beliefs about how best to help students thrive? It's complicated. Two key aspects for consideration on this front are (1) unexamined assumptions and (2) the opportunity gap. We will devote the remainder of this chapter to better understanding both aspects.

Examining Assumptions

Assumptions, *implicit bias*, and *unconscious bias* describe the ideas or internalized beliefs impacting the decisions we make, often without ever registering in our conscious thinking patterns (Hammond, 2015). It can be uncomfortable when these weak spots are brought to our attention. I liken this exposure to the experience of driving a car and having a passenger urgently alert you to something you may not see. "Watch out; there's a cyclist!" my passenger shouts. I slam on the brakes instinctively, whip my head to see what is there, and then say, "Jeez! You don't have to shout!" What I *mean* to say is, "Thank you," but my initial reactions are usually in defense of my ego and a reflection of my eagerness to shift the attention away from *what I failed to see*—from my mistake. I can confidently say I mean to be a conscientious, alert, and attentive driver. As drivers, though, sometimes we all just *miss* something. For me, it takes practice and humility to course-correct when I am doing the best I can already. At the same time, it is a necessary aspect of continued growth, both personally and professionally.

When left unexamined, our assumptions can lead to decision making that risks perpetuating the systems and practices that marginalize our students and colleagues. As we confront the tension between what we *intend* to have happen as a result of our decisions and the actual *impact* of those decisions, it is wise to practice thoughtful reflection about our assumptions and carefully check in with ourselves. Attending to the nuances of grading policies helps us focus on the impact of those policies without investing too much time defending their intent. Through continuous excavation of our *unconscious drivers*—our weak spots—we position ourselves to better align our next choices with our desired impact. In my case, this work unfolded as I examined my role in perpetuating or disrupting my district's grading system. There are two areas I encourage you to self-examine as well: (1) checking your weak spots and (2) critiquing your practices.

CHECK YOUR WEAK SPOTS

My school's established grading policy had teachers generate grades based on participation and classwork, homework, and assessment. It is useful to unpack each of these components in detail, considering each line item through the lens of who it *helps* and who it *harms*. Figure 6.1 (page 104) provides one useful framework for examining your areas of unconscious bias. There is space in each section to add your thoughts. Your unique experiences will afford you a perspective that might be different from mine.

Initially, I did not see the assumptions about students inherent in this approach to grading. It took a few semesters operating within the system to see how the established structure for determining student achievement did not fairly represent each of my students' efforts to learn and may have been contributing to the cyclical *disengagement* many students experienced.

Participation and classwork: ***Typically group work, showing your work on the board, volunteering an answer, and so on***	
Who might this help?	**Who might it harm?**
Students with robust mathematical identities Students with a favorable relationship with the teacher Students the teacher notices Students comfortable in class Students comfortable sharing their ideas Students able to show up physically and academically that day What else would you add?	Students with fragile mathematical identities Students perceived as disruptive Students who do not get to school or class Students uncomfortable interacting with their peers Students requiring differentiated opportunities to share their ideas What else would you add?

Homework: ***As traditionally assigned, such as page 147, problems 1–23***	
Who might this help?	**Who might it harm?**
Students with time and space at home for schoolwork Students with enough understanding of the mathematics so practice is useful Students without competing time demands such as work, after-school engagements, family commitments, and so on What else would you add?	Students with partially developed understanding (the repeated practice of errors can create a greater learning challenge to overcome) Students without time and space to do schoolwork outside school Students with class loads that require hours of nightly homework What else would you add?

Assessments: *Such as typical midchapter quizzes and chapter tests*	
Who might this help?	**Who might it harm?**
Students in class that day Students with an opportunity to study and prepare Well-rested, well-fed students What else would you add?	Students who have missed a good deal of in-class instruction Students without opportunities to study outside school Students encountering various environmental stressors What else would you add?

Figure 6.1: A framework for examining your assumptions.

Visit ***go.SolutionTree.com/mathematics*** *for a free reproducible version of this figure.*

CRITIQUE YOUR PRACTICES

Use critical questions to examine your assumptions, checking to see how any given teaching choice helps or harms. It is crucial to engage in reflective practice, intentionally inviting into the discussion those colleagues or skilled practitioners who may shed light on something we are not already attending to. These collaborative relationships inform the work, often recommending more interesting questions to consider as we continue to learn and grow. While difficult at times, strive to embrace the discomfort of your mistakes. There is a lot of information to glean from those mistakes, and we do well not to shy away from them. It is hard and uncomfortable, but worth it. The following are questions to use as you begin to dive into the assumptions made in your (or your school's) grading policies.

- How does this participation-points approach take into account the diverse reasons why students may miss class on any given day?

- How does this approach create a self-promoting system that values compliance over engagement?
- Why are we determining *how* a student participates as a measure of learning?
- Does the narrowly defined *participation* category include *multiple ways* a student can engage in learning and demonstrate understanding?
- How might this approach create a greater divide between well-resourced students and less-resourced students?
- How might the grading structure need to change for it to provide meaningful information to students about their learning journeys?

More broadly, use the following questions when you aim to better understand your own assumptions.

- If I keep on this path, who will it help? Who will it harm?
- Who is missing from this conversation? What voices, cultures, experiences, and perspectives am I not considering?
- Whose experiences am I considering in this situation? Who do I include?
- Who am I believing? Who am I questioning? Why?
- Who can help me see what I do not even know I am missing? Who can check my weak spots?

This leads to an all-important question: If not the traditional grading structure, then *what*? Teachers and leaders continue to experiment with various approaches, including eliminating grades altogether. What? No grades? You can probably hear it now. Adults at home, teachers, students, and administrators, all with their own concerns about what it would look like or mean to not have letter grades. "How will we know?" they may ask incredulously. Listen for the unasked question within that question. How will we know *what*, exactly? What are we hoping to know about students from their grades? Do grades actually deliver important understanding for students, their adults at home, teachers, or administrators? If grades do not deliver that information, then what will? Let's start with instructional routines in our classrooms, then zoom out to consider schoolwide and systemwide implications.

IN OUR CLASSROOMS: DEVELOP BETTER ASSESSMENTS

As teachers, we always hope grades accurately reflect student learning and inform students of their progress toward specific learning goals. It seems simple enough. It certainly seems well intentioned. We must ask though, "Does an *A* do that? Does a *C* do that? Or an *F*?" The answer is, it depends. Students are often hungry for affirmation in the form of an *A*, or crushed by rejection in the form of an *F*. Because we aim to "move the needle" on engagement and belonging, we must shift our grading practices over time to be more transparent and provide clearer success criteria, paired with more meaningful feedback (Hattie et al., 2017).

To make feedback meaningful, post descriptions about what it means to understand a particular concept or skill the standards designate students are learning, and what students will learn and do next as they progress toward that target (Hammond, 2015; Hattie et al., 2017). As a result of effective feedback, students' metacognitive and self-regulatory skills improve (Hattie et al., 2017). You will hear students transition from saying things like, "What are my missing assignments?" to "I don't understand how to show how to divide fractions." The outpouring of the shift on my part was an intentional focus on *what we were learning*.

You may be noticing or even concerned about (as I was) how participation in conventional grading constructs inadvertently perpetuates cycles that create learning dependency on factors that felt arbitrary at best and preferential at worst. By *learning dependency*, I mean students feeling proud of their achievements, perseverance, and efforts, and then exhibiting feelings of dejection and exasperation when they read their report cards. They could not reconcile the gap between how hard they were trying and their grades. Neither could I. Hear me when I say, "I am not suggesting everyone get *A*s for trying hard!" I envision a more elegant system—one that focuses on whole-student, personalized growth. A system that provides meaningful feedback to help students know their next steps in learning and contributes to their learning journeys, rather than obstructing them.

When it comes to shifting from *grading*, keep in mind the importance of *correcting*. Receiving feedback we can use to improve understanding and performance fuels motivation and engagement (Hammond, 2015). Getting a letter grade or even a total score of 4/10 is not informative for students; this is not helping them know where their learning is solid and where it can go next.

Let's use test corrections to illustrate how to shift from just providing a score to providing useful feedback. To begin, correct student tests and indicate which problems they correctly solved and which they need to revisit. Then, use the following process to engage students in instructive (rather than evaluative) feedback (Hammond, 2015).

1. For the class period following the assessment, students arrive to see each test question from the day before posted on sheets of paper around the room, one question per page. (Students who were absent on assessment day receive a copy of the test to complete.)
2. Distribute tests back to students and provide about five minutes for them to review your comments, as well as their own work. Provide time for students to revisit and resolve any problem they felt they could tackle without assistance. Consider using mathematics journals or spiral notebooks so they have plenty of room to work.
3. After individual correction time, students take their journals (or spiral notebooks) to the test question paper posted in the room that matches a test question they had not correctly solved.
4. At the bottom of each posted paper, write one or two student names as "experts" for students who missed the question to consult for help. Students might read something like, "Check in with Viera for help! She's a pro at this!"

In preparation for this test-correction experience, review the tests of all students—even those who only got two or three problems correct—and ensure each student is named an "expert" on one of the problems, which is an example of *assigning competence* (Cohen, Lotan, Scarloss, & Arellano, 1999). I endeavored to distribute the knowledge authority among all my students, publicly elevating their thinking by identifying them as the "experts" and thereby cultivating a reciprocal sense of belonging and engagement (Hammond, 2015).

During the test-correction activity, students circulate, checking in with one another when they need an "expert." You can determine the amount of time to allocate for this. Circulate while students work, listening to their discussions for any common themes that may require whole-class discussion. You can gain valuable information about calculation errors versus conceptual understanding by listening to student discussions.

While some educators pride themselves on tough testing, I am not one of them. If the entire class misses a test question, then something about the learning experiences on that concept or skill must not have met student needs, or the test question itself may be off. Either way, it warrants investigation. To prepare for reteaching, ask students to describe in their journals what had been tough about a particular missed question; this information will help you know where they got stuck or confused. We need students' stories and narratives to better understand how to design future learning to effectively convey the key concepts.

PAUSE AND PONDER

How does this experience of assessment, feedback, and learning compare to conventional approaches? As a teacher, how does it feel? How do you think the students felt?

When shifting from a focus on a single assessment to the broader scope of reporting on student learning, consider the multiple measures of student understanding of the content standards they are learning. Compile data from in-class work and assessments that align to student learning of certain standards. Create a conversion calculation that weighs those standards based on importance and generate grades from there. For any student whose grade converts to a *D* or an *F*, I encourage you to enter *INC* (incomplete) or the equivalent, so as not to inadvertently perpetuate student tracking. This didn't always go well for me (as I described earlier, see page 100). And although we may encounter friction as we shift our professional practice, it is a necessary part of our work to align grading practices with our other efforts to cultivate student engagement and belonging.

IN OUR SCHOOLS: QUESTION CONVENTIONAL GRADING

When we find ourselves more allegiant to a system than to the people in that system, it is time to start asking questions. Allegiance to the system can sound like the following.

- "It's too late to turn in that assignment" or "I do not accept late work."
- "You can make test corrections, but it won't change your grade."
- "Late assignments are marked 10 percent down."
- "Forgetting to put a name on your paper will cost you 10 percent."

The opportunity gap is the second factor crucial to consider as we advocate for equitable grading systems reflective of our commitment to student success. While many of us are steeped in discussions about the achievement gap, the opportunity gap takes our thinking in a more useful direction.

The term *achievement gap*, as you likely know, describes the gaps in achievement between White males and everyone else standardized tests measure (NCTM, 2018). Pay a quick visit to the National Assessment of Educational Progress (NAEP) website, and reports, data, and metrics will overwhelm you (see https://nces.ed.gov/nationsreportcard). These are *effect* data. All the gathered data is a summary of the outcomes of assessment across the United States. You may also notice options for viewing the data are to compare females to males, Hispanic to White, Black to White, and so on. These effect data report the achievement of White males in our education system outpaces the achievement of other students. However, conversations that center on the achievement gap do not invite us to ask the more interesting questions, such as the following.

- "What systems are in place that support the achievement of White males?"
- "What systems are in place that make it more difficult for other students to thrive?"
- "How are we assessing students? Who writes the tests?"
- "What role does language play in mathematics assessments?"

These questions (and about a zillion more) aim at what *causes* the data. This is the beautiful, impactful, and complex conversation about the opportunity gap. Welcome! Let's get into it, starting with deepening our understanding of the opportunity gap discourse.

IN OUR SYSTEMS: ATTEND TO THE OPPORTUNITY GAP

The *opportunity gap* highlights the inequitable distribution of resources and opportunities among groups of students, which demands our attention (Flores, 2007; Gutiérrez, 2008; Ladson-Billings, 2006, 2014; Paris, 2012). Affording differentiated opportunities to different groups of students may be evident along various categories including race, ethnicity, place of birth, socioeconomic situation, and more. These circumstances influence peoples' paths and often influence if and where they go to school, which in turn influences the types of learning opportunities they're provided. There are broader, systemic consequences for our practice as professional educators, including involvement in policy discussions, districtwide implementations, and political advocacy. Initially, however, we can focus our attention and efforts on what we can immediately control in our own classrooms, and discussions we can initiate with our colleagues.

If your experience is anything like mine, I suspect you regularly engage in conversations about student data, including with some form of data boards and focus teams. *Data discussions* often include analyzing the performance and achievement of student subgroups. It pains me to say (but it is also important to tell the whole story) that in my school, these data discussions often included strategies to target *bubble kids*, which were the students just on the cusp of the state's *Meets Standards* threshold. We did not realize how our conversations about students and their achievements on standardized tests reduced them to scores and data points. While we intended to figure out which students needed additional learning opportunities, these conversations did not bring the answers we needed.

Repeated discussions about the *achievement gap* hindered our ability to see the *opportunity gap*. We fell short of critiquing the systems in which we asked our students to perform. In efforts to show student achievement, we inadvertently lost sight of the humans we aspired to teach, settling into the system we thought was measuring how well we were teaching them. At some point, we began to realize there was much more to the story; we were missing the key insights into student learning that would ultimately be a fruitful place to expend our energy.

Grounded theory researcher Brené Brown (2022a) describes *regret* as either a failure of courage or a failure of kindness. My regret about the data conversations includes both. This may be part of your journey as well; integrating improved understanding into your practice may mean acknowledging the shortcomings of previous endeavors. It is wise to spend a little time with any emotions that surface as a result. I encourage you to bring compassion with you on that journey. We are all continually learning more so we can keep doing better. Wherever you may be along this trajectory, lean into the power and potential of engaging with the complexity of each student's developing and multiple identities and not reducing each student's learning to constrained, biased evaluations.

In *Catalyzing Change in High School Mathematics*, the NCTM (2018) summarizes what we all have seen unfold time and time again:

> Once students are placed in a low track, moving out of that track is very difficult. Tracking students into weak, low-quality, and dead-end course pathways creates a gap that widens as it puts students further and further behind in the curriculum, leading to what is commonly referred to as the "achievement gap." This gap is more accurately labeled as an opportunity gap, since it is largely manufactured in schools by practices that place students into low-level courses and dead-end learning pathways, often with less effective instruction. (p. 17)

This understanding is at the root of why I knew it would be so detrimental to my students to translate their *INC*s into *F*s.

As we develop more keen interpretation skills, we start asking critical questions aimed at closing the opportunity gap. As you consider your students' learning opportunities and your opportunities to address and meet those needs, engage your colleagues, administrators, and central offices staff in discussing the following.

- Providing translations for mathematics assessments
- Providing instructional materials and teacher communications in students' home languages
- Addressing the disparate quality of learning students experience from teacher to teacher and class to class

The more questions we ask, the more evidence we can gather about the nature and impact of our established grading practices. Discussions and questions on this subject may make your colleagues uncomfortable; I know because this was my experience as well. Stay curious about what motivates colleagues to prioritize allegiance to the school systems, rather than the people in the school. Think with me for a moment: What if, at your next department meeting, you asked to dive deeper into the opportunity gap rather than fixating on the achievement gap? How might your colleagues receive that question?

Dedicate reflective practice and professional conversations to the multiple factors composing the opportunity gap. Seek opportunities to engage in critical discourse about teacher tracking. Critique access to school, like the systems that make it easier for some students to get to school than others. Collaborate with colleagues across your campus and district to make change. Examine your specific students' access to learning, such as what quality curriculum or materials you leverage in designing lessons. Throughout all discussions, surface and address equity. Whatever the reasons for opportunity gaps are for your students, keep talking about them. Keep surfacing the *cause* data rather than being swept up in discussions about the *effect* data. Use those conversations to deepen conceptual understanding and determine strategic action to redress inequity. As we work to mitigate the causes of disparate educational experiences, the achievement gap will take care of itself.

To Sum It Up

Recognize the long-term implications of Fs and your role in closing the opportunity gap.

At the risk of sounding trite, assessment is complicated. The conflict with my principal was not solely about my performance as a teacher and the administrative system in which I operated. It was not only about those few students and their achievement according to the system's standards. It was just not that simple. As a group of dedicated professionals, let us keep having the complicated, necessary conversations that push us toward more equitable, more just systems that accurately represent the complexities of teaching and learning.

Remember the two sentiments from the beginning of this chapter (see page 99)? Who is responsible for the *F*, the teacher or the student? We are all responsible for teaching and learning, and it will take each of us to develop a more refined system of correction and feedback that meets student learning needs. Conventional grading simply does not. Conventional grades say, "Here is where you measure up in a system that may or may not be designed for your benefit." When students receive failing grades in this system, they are then tracked into low-ability groups and terminal course trajectories. It does more harm than good. Recall the process I used the day after an exam? After the test-corrections class period I described earlier (see page 108), a student approached me and asked, "What is my grade? I only got 4 of all 15 right. Is that an *F*?"

"Nope. It means there are some concepts you totally understand and others you are still learning about. And it means I need to take a closer look at some of those test questions."

"But like, that's an *F*, right? Why didn't you just give me an *F*?"

"I am all out of *F*s, sir. I have no *F*s to give," I smiled, and he looked noticeably relieved. This was a new beginning for this eighth grader, a student in mathematics intervention who had experienced cycle after cycle of *F*s. That is not what we do around here. *F*s on papers interfere with engagement and belonging, so there are no *F*s in Portable 137.

Chapter 6 Application Guide

Use the reproducible "Chapter 6 Application Guide" to gather your thoughts and make connections between the themes and tools in chapter 6 and your teaching practice. Consider using this application guide collaboratively with colleagues to anchor your exploration, discussions, and application of your learning.

Chapter 6 Application Guide

Chapter Themes and Tools	Connect and Apply to Your Practice
Examining assumptions	Choose a practice area (such as assessment practices or grading) and engage in a "Who does it help and who does it harm?" analysis. Consider engaging colleagues in a discussion to broaden your perspective by hearing ideas that may not be on your radar.
Developing better assessment and grading practices	Consider an upcoming lesson or assessment. How might you engage students in revisiting and correcting errors? How might you measure the impact of your approach on their attitudes?
Attending to the opportunity gap	Learn more about the trouble with the achievement gap (see recommendations in appendix A, page 155), and consider how you might share your learning with colleagues or building and district leaders, shifting focus to the opportunity gap.

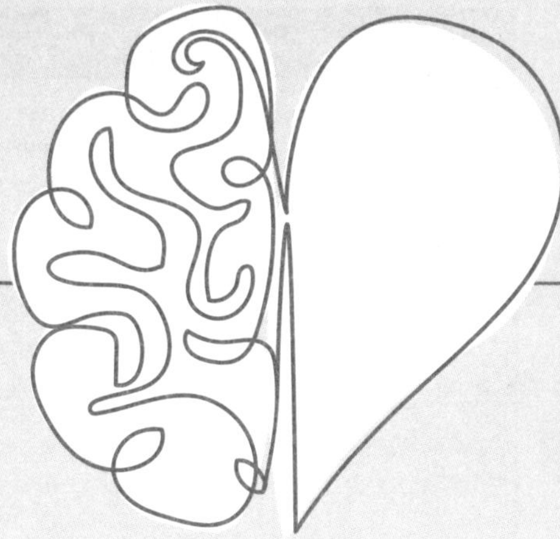

Don't let what you can't do

interfere with what you can do.

—John R. Wooden

CHAPTER 7

If Only Students Knew Their Facts

During our midweek teacher planning meetings, our mathematics teacher team often collaborated on upcoming lesson planning. The work often included diving into student results on common formative assessments, discussing upcoming units of instruction, and creating the lessons the team deemed necessary to preteach for students to succeed with upcoming content. There's a lot of collective wisdom we can glean from co-planning sessions such as these, and, for me, these sessions included the opportunity to hear from core mathematics teachers about students' learning progress. Consider the following.

"Oof. Next week, we start simplifying expressions and combining like terms. Not looking forward to that," a colleague expressed. "They're just not ready."

"I know! Those topics can be so brutal for kids, especially those who don't know their facts," another added.

I side-squinted a little. "What do you mean?" I asked, authentically unclear about the conclusions they were drawing.

"They're just so slow! They can't quickly add or subtract, much less multiply. They have all this trouble understanding the new stuff because they haven't even memorized the basics! They just quit; it's too much to figure out. And we don't have time to teach them the facts!"

Aah, I thought. Interesting. Also, I wondered to myself, "Just because a student knows the facts, is the student necessarily ready for different types of learning?" I felt internal resistance to the logic I was hearing, yet kept my ideas to myself, based on the social structure that values years of experience (of which I had few). I left the discussion with new questions. This was my own matter of belonging to navigate as I endeavored to engage in the co-planning discussion. Let's look further at limiting sentiments such as, "If they only knew their facts," and strategically unpack and reframe our own narratives (building our conceptual understanding) so we can craft learning experiences that engage students at various levels of readiness.

What's Really Going on Here

By the time students reach mathematics in middle and high school, several compounding factors are at play. Traditional approaches to teaching mathematics become increasingly abstract and often fall short of including learning opportunities dedicated to the sensemaking required for deep, lasting understanding to take root. We also have engagement shouting for our attention. Poring over preassessments to try to discern what students are ready to learn often reveals how much more information we need before we feel prepared to map a course of learning that makes sense.

As you have likely also experienced, while some of the assessment questions helped me know the students' landscape of understanding, others left me wondering, "What does this right answer tell me about what they understand? What does a wrong answer tell me?" Our formal preassessments typically capture the thinking students chose to convey on their papers, and teachers understand that students' thinking is not effectively captured in this format. As we begin to think about the bridge between gathering data and actually writing lesson plans, your mind may be swirling —as mine often was, trying to think about the conceptual mathematical underpinnings across grades 6, 7, and 8. How *would* I differentiate instruction to meet student needs? How *would* I help an eighth grader with number sense without creating a sense of doing "little kid" mathematics? We'll dive into strategies, but before we do, let's examine a counterpart to these content-engagement wonderings: *belonging*.

Not only does secondary mathematical content become more difficult for teachers and students but also the social dynamics of learning and asking for help grow increasingly difficult as students approach and navigate adolescence. Matters of belonging demand our focus. Ever more interested in peer approval, students developing multiple and complex identities may not respond well to teacher affirmations or approval (it's not "cool" to be good in school). Furthermore, many students carry additional responsibilities outside school. Some students drive or ride the bus after school to the elementary schools to pick up younger siblings before heading home and shouldering the responsibility for most caregiving tasks. Answering a question such as, "How can I teach effectively?" becomes nuanced and complicated given the realities of students' lives. The challenges here are *real*.

Radical Acceptance

I was first introduced to the concept of *radical acceptance* while working with a therapist, and since then continue to work on developing my skills to accept the reality before me, rather than debating it or wishing it were not the case. "Radical acceptance means practicing a conscious effort to acknowledge and honor difficult situations and emotions. Fully accepting things as they are, instead of ignoring, avoiding, or wishing the situation were different" (Brown, 2022b). No debates, no *what-ifs*, just acceptance that *what is*. Because we are human and bring our whole selves to our classrooms, the roots of this dialectical behavioral therapy practice are relevant to our classroom strategies. Like many facets of my growth, this is a work in progress. *Acceptance* played a key role in my ability to move into creative problem-solving space when it came to finding lessons and approaches that would meet students where they were. Without bringing expectations or an unspoken desire that students knew more than they did, I chose to radically accept reality.

Students are incredibly perceptive. We don't have to *say* anything for them to *sense* our frustration with their lack of fact fluency or challenges with conceptual understanding. We also know the power of creating learning environments where students don't have to pretend to know more than they do or hide their confusion. As teachers, we need to get this out in the open to know what students *really* understand (and what they don't) to make an actual difference in mathematics learning. Rather than bemoan students' partially developed understanding, the premise of *radical acceptance* is to accept reality and move on from there. Rather than saying, "If only. . .," adopt the internal rephrasing of, "Because . . .," as in the following.

- From *If only they knew their facts . . .* to *Because they do not know their facts . . .*
- From *If only their parents came to conferences . . .* to *Because their parents were not at conferences . . .*

- From *If only they did their homework . . .* to *Because they do not do their homework . . .*

There is a responsibility shift in these statements, a blame shift even. *If only* places the responsibility on others (students, parents, last year's teachers). This perpetuates the internal story that *I cannot do my best work because of others*. What a frustrating, powerless position to be in. Turn that unproductive narrative on its head! *Because* puts the ball in your court (and that of your colleagues). It radically accepts what is *real* and poses the question back: *What am I going to do with this reality?* This does not diminish the factors outside your control. Those factors matter. But many factors are also outside the student's control. Seek to disrupt deficit-focused thought patterns by choosing to adopt more productive ones that say, "This reality is tough" and "This reality doesn't have the final say in whether I can make a difference here."

Taking this approach can help you stay present to the needs of students, regularly hold the door open for their realities, and remain unshocked by whatever new challenges arise. This perspective can help you keep an open mind, always scanning the landscape for what is possible to achieve given the constant influx of new obstacles. With this open mind, you are in a better position to be a creative problem solver, to think beyond the constraints, and to approach each day and each lesson responsively.

PAUSE AND PONDER

Where might a radical acceptance of reality help you move from feeling stuck to feeling like your efforts make a difference?

When we operate from a place of radical acceptance, we can transfer our energy from analyzing what we can't control to what we can. This not only diminishes the potential negative momentum of our daily experiences but also demonstrates a belief in what the data show; this year, with this student, your efforts matter. External factors do not matter more than the work you do with students—the high-quality lessons and the opportunities to grow that happen on your watch.

Collective Teacher Efficacy

John Hattie and colleagues (2017), in research about influences on student achievement, identify a concept they call *collective teacher efficacy* (CTE). According to Hattie's research:

Collective Teacher Efficacy (CTE) is the collective belief of the staff of a school/faculty in their ability to positively affect students. CTE has been found to be strongly, positively correlated with student achievement. A school staff that believes it can collectively accomplish great things is vital for the health of a school and if they believe they can make a positive difference then they very likely will. (as cited in Visible Learning, n.d.)

Hattie and colleagues (2017) measure the impact of this concept using *effect size*, which "represents the magnitude of the impact that a given approach has" (p. 20). CTE has an extraordinary effect size of 1.57 (Hattie et al., 2017). For reference, Hattie's research shows factors that significantly impact student learning have an effect size of 0.4 or greater (as cited in Visible Learning, n.d.). To see where CTE stacks up, consider that class size has an effect size of 0.21 and classroom discourse is 0.82. CTE is *three times* more predictive of student achievement than socioeconomic status. When it comes to student achievement, it's not that other factors *do not* matter; it's that you matter *more*.

When teachers believe they matter, they make decisions accordingly. Teachers who believe in their collective efficacy demonstrate a willingness to try new approaches. They set more challenging goals, and they attend more closely to the needs of students who require extra support (Hattie et al., 2017). With this in mind, let's explore two major shifts you (along with others in your school) might make to truly impact the learning trajectory for your students: (1) reconsidering homework and (2) investigating underlying mathematical concepts.

RECONSIDER HOMEWORK

Here is the deal: learning in class cannot depend on the previous night's homework. It doesn't work, and it's not equitable. Setting up learning with these types of prerequisites generates cyclical, unnecessary struggles for you and your students, and drives disengagement. I learned this lesson the hard way, initially following a traditional approach to assigning a little homework and beginning the next class by addressing questions or concerns students may have. I thought I had an innovative approach to this traditional practice. Homework assignments were creative (so I thought), and we did homework corrections collaboratively. Students would come to class and write their challenging homework problems on the board, and other students would solve them and explain their thinking to their peers. Even this, however, poses equity and access concerns, and takes too much class time. I ultimately needed higher-impact learning opportunities. Realizing this caused me to question the effectiveness or value of homework, and once I did that little bit of radical acceptance, everything changed.

It is important to examine any well-intentioned practice (such as homework) to consider the role of that practice in student learning. In my case, homework became divisive (between students who *did* their work and those who *did not*) rather than unifying. It also derailed lesson plans because I made the mistake of planning the next day's lesson based on factors outside my control, such as whether students completed their homework the night before. It was problematic on many fronts, so I stopped giving homework. An empathetic, perspective-taking moment also helped me see how harmful this practice was to students in intervention who were working to recover from years of frustration learning mathematics. A student walks into class, and if the student was absent last time or didn't do the homework, the experience the student has in the first ten minutes of class is one where the student cannot possibly succeed. This is not an accessible, responsive approach, and it only reinforces the disengagement and unbelonging the student faces, rather than disrupt them.

So what do you do instead of assigning homework? Without homework, you will likely recover minutes of class time. You can reallocate this portion of learning time to learning experiences that do drive engagement, belonging, and, ultimately, student learning. It means more time to do the following, for example.

- Launch lessons with engaging, lively discussions designed to foster number sense.
- Process learning at the end of a lesson.
- Investigate mathematics during a lesson.
- Engage in mathematics discussions.
- Attend to social-emotional needs throughout learning.

Not only can you recover important instructional time but these shifts are also wise investments in building confidence and a love of learning, which is better for you and students. Win, win, win!

PAUSE AND PONDER

Where does the success or failure of learning currently rely on something out of your control, and how might you reimagine that practice?

Even if you are not part of a collaborative team already, don't be surprised when the impact of this approach spreads beyond your classroom.

"How do you correct homework?" a colleague asked me during one of our department meetings.

"I don't assign homework." I was the only teacher on our team who did not, and the opportunity to build a little CTE emerged.

"Well, that makes sense. Your students have mathematics homework from their other mathematics classes. They don't do it anyway, but it makes sense. Maybe they could use some of their intervention class time to complete their homework from our classes." This demonstrates a common misunderstanding about what *intervention* is and which teachers are responsible for what aspects of students' core instruction.

"Well," I began as I took a deep breath. "I don't assign homework because it doesn't help most students learn. If you already understand mathematics, then homework doesn't help you understand it more. Conversely, if you do not understand it, then homework also does not help you learn it. It is not equitable, as many of our students have out-of-school routines and responsibilities that we should allow them to prioritize. I think there's a better way."

This comment from me was met with silence, pondering faces, furrowed brows, and disequilibrium.

I wasn't likely to sum up this discussion in a single meeting. (You will hopefully revisit any discussion like this as many times as needed to deepen collective understanding and shift instructional practices.) In my case, we had the "What do you gain or lose?" conversation repeatedly. We encountered challenges with our school's grading policy, which included teachers calculating a percentage of the grade based on homework. Our process was not fast, nor was it straightforward, but through slow-earned collegial buy-in, we persevered. Many on this team wanted to help students learn and were willing to try new approaches in that pursuit. By the time I moved on from teaching at that school, the mathematics department had shifted its homework policy to be more inclusive and student centered. The department hadn't fully departed from homework, but assignments started to sound more like the following.

- "Between now and our next class, see if you notice any hexagons out in the world."
- "Between now and our next class, notice how many percentages you see in ads or on signs."
- "Between now and our next class, think about situations with input-output relationships."

This approach results in students thinking about mathematics and noticing the world around them. It results in teachers learning more about the kinds of things students notice and the perspective each brings to class. It results in next-class discussions in which all students participate. It removes barriers to learning and reduces tensions. For me, it meant no longer did our department meeting discussions focus on the statement, *If only they did their homework. . . .* Instead, our energy focused on filling in the blanks after *Because homework doesn't work . . . .*

INVESTIGATE THE UNDERLYING MATHEMATICAL CONCEPTS

Recall that sixth, seventh, and eighth graders were intermixed in my intervention classes. The instructional implications necessitated I find the conceptual underpinnings common to the standards across several grade levels. It meant structuring learning tasks that gave students opportunities to revisit and solidify partially developed understandings on which grade-level learning depend. To find the common threads, examine the data about what standards a student has not yet mastered and work backward to determine the conceptual origins of the topic. What are the conceptual roots of this student's struggle with a seventh-grade standard about expressions and equations, and another student's struggle with a sixth-grade standard about dividing fractions? Let's explore, shall we?

An often-heard request from a student's core mathematics teacher is, "She needs help with [a skill]." This sounds a lot like a patient arriving at a doctor's office and describing symptoms. While symptom management is important, if looking to achieve long-term improvements in a person's overall health, a physician must investigate and address the *causes* of those symptoms. The same is true for teaching mathematics. While students demonstrate difficulties with certain skills, simply addressing those skills is insufficient. We must get at and solidify the conceptual understanding that leads to fluency with procedures. When we address the concepts (causes), the skills (symptoms) will often resolve themselves. Realizing our students do not know their mathematics facts fluently (symptom), drill-and-kill practice or tricks for memorization fall short of providing the opportunity to develop deep conceptual understanding from which fact fluency will naturally stem (Bay-Williams & SanGiovanni, 2021; NCTM, 2014).

Understanding the relationship between concepts and skills matures over time, and may lead you to focus units of instruction or sequences of lessons on things like *finding and describing patterns* or *building number sense.* Across the grades, the types of patterns students analyze or the kinds of numbers they work with might be different. However, the *key concepts* of noticing and describing patterns using multiple representations (pictures, tables, words, graphs, and so on) are consistent.

As you plan lessons, know where the standards come from and where they are headed to understand the trajectory of student learning. It seems a more routine practice is to preview assessments and upcoming learning to know where lessons aim to take the learning. It is equally, if not more important, to know where those standards *come from*. Radically accepting that many students are not yet prepared for on-grade-level mathematics requires us to plan instruction that intentionally revisits and often shores up understanding of prior grade-level content. Learning the progression of key domains—such as number sense and operations in base-ten—from the very beginning (yes, I mean preK and kindergarten) is important. If we start to explore the sources and depth of a student's confusion way back in first grade, then we need to be prepared to engage that student with the learning needed without the shame or expectation the student should understand something the student simply does not. This may seem daunting. Be encouraged; it is doable.

Let's use number sense and operations in base-ten as an example. When secondary students struggle to factor polynomials, we can investigate how well they understand concepts about factoring more broadly. We can dig into factors and multiplication, explore decomposing and composing whole numbers, and so on, as we pursue the source of confusion to its origins and begin solidifying concepts from there. "My students just don't have any number sense" is another statement I frequently hear. *Because* students need help developing number sense, *we* must know where and how to develop number sense to meet their needs and move them forward.

LEARN FROM COLLEAGUES

As you delve into prior content, conversations with elementary teachers can help you better understand mathematical content you are less frequently tasked with teaching. We can learn from teachers of elementary grades how they approach developing notions of a *base-ten number system*, which is the unit on which our place-value system is built. It was through discussion with elementary teachers I first learned about using dot images, ten-frames, and the hundred grid to help students build number sense. In studying the content standards, I began to realize the conceptual underpinnings of how mathematical operations work with *any* number are essential background knowledge for students to successfully operate with fractions. How do you know when this is necessary? You may find students can perform a particular calculation or execute an algorithm but are unable to explain the *why* behind those calculations.

For my students, number sense and operations were *the* most important content to revisit in mathematics intervention, and engaging students in learning experiences targeting those concepts unlocked countless skills. Figure 7.1 includes a few of my favorite learning routines related to number sense and operations. As you review these routines, think about underlying concepts your students may need opportunities to learn, and how you might structure those opportunities to cultivate the common denominators of engagement and belonging.

As you plan learning activities such as those in figure 7.1, you may hear or even feel your own sense of concern about doing something like a dot talk with secondary students. I've heard the concern, "Wouldn't students feel like they're being treated like little kids?" but in my experience, they don't. The difference was the *rich* conversation stemming from these dot talks, plus students getting the chance to make sense of something, share their thinking, and connect their approach to other students' ideas.

Dot Talks, Ten-Frames, and Number Talks	
Dot Talk	
What	Create a set of cards, each with an array of dots. Choose one card to start. Show the card to students for about three seconds and then remove it from the set. Then ask students, "What did you see and how did you see it?" Students share what they saw, describing how they saw it as well as the total number of dots. Importantly, write students' ideas either on the board or on chart paper to create a record of their thinking, ensuring their names are next to the strategies they share. Additionally, correlate number sentences to the word sentences students share to show how their ways of thinking *are* mathematical. Discussions about student ways of seeing the arrays also lead to important conversations about equivalence and the properties of operations.
When and Why	Use this activity as a warm-up to get students thinking and talking about what they notice. This low-stakes opportunity for learning helps students experience quick wins while solidifying their skills to flexibly compose and decompose numbers, formally known as *subitizing*. Eventually, you can extend strategies for seeing arrays of dots to seeing *arrangements of dots* in ten-frames. This supports familiarity and practice with the base-ten number system and composing and decomposing numbers to find sums or differences.

	For example: 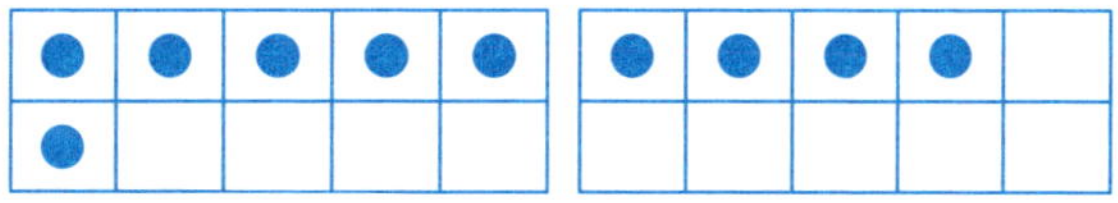 This activity ultimately leads to number talks using arrays of images that help students identify patterns that support discussion of properties of operations (Parrish, 2022). Following is a series of dot images I used for one of these discussions. 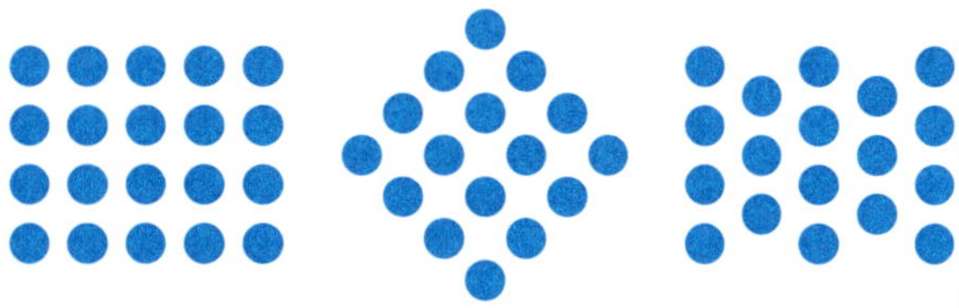
Make Ten	
What	Roll a handful of dice and ask students to find all the possible combinations to make ten.
When and Why	Use this game as a warm-up, a learning break, or an optional game for students to play during one of their learning rotations (engagement). By removing the pressure (belonging) to "add them all" or to "get the right answer fast," this activity prioritizes multiple solutions and creative thinking while simultaneously providing students an opportunity to practice looking for groups of ten, a key landmark in developing procedural and computational fluency.
Number Puzzles or Equation Puzzles	
What	Students receive a set of cards with numbers and operation signs written on each card. In teams, students work to create an equation that is true using the numbers and signs. They don't have to use all the cards in the set to achieve this. Create the cards so students can use more than one possible arrangement. For example, an initial set of cards might include all four operations, an equal sign, and the numbers *3*, *4*, *8*, *12*, and *24*. Depending on students' comfort and understanding, consider starting with whole numbers, eventually adding integers, rational numbers, variables, and inequalities to the mix.
When and Why	Use this activity during collaborative learning stations (belonging). Differentiate card sets based on students' conceptual understanding (engagement). The purpose of this activity is to focus on the idea of equivalence as well as flexibility with number sense and operations.

Figure 7.1: Routines for developing number sense and operations.

Visit ***go.SolutionTree.com/mathematics*** *for a free reproducible version of this figure.*

When it comes to cultivating engagement and belonging, this experience is crucial. Just because students are of a certain age does not necessarily mean they have solidified understandings of number sense and operations. In fact, many are developing skills to mask their confusion as they progress through school. Just because students can recite the definition of the *commutative property of addition* does not necessarily mean they understand *why* that property works or *how* that property might help them think about flexible and accurate calculations with variables. Targeting the root of the confusion is a worthwhile time investment. My students showed tremendous growth as a result of this approach, and so can yours. Allow me to provide details to illustrate how to make the most of your growth-monitoring efforts.

MONITOR GROWTH

I analyzed student growth data each year, and because the district used state test scores to determine intervention placement (an example of student tracking), I also used state test scores as a summative assessment of student learning. Although these standardized assessments were imperfect, they were the metrics the district used throughout achievement and needs discussions, so I calibrated my language accordingly. Our data showed students eligible for and receiving mathematics-intervention instruction showed an average of nine to ten times the growth of their peers who were also eligible for intervention, but did not receive it. This makes sense, given the significant influence of response to intervention on student learning (Hattie et al., 2017). Gathering these data is important as teachers and administrators continue to advocate for and justify the expense of mathematics-intervention resources, including salaries and teaching positions.

Equally if not more compelling than state test–score data are the qualitative data we gather as students reorient themselves to their own multiple intelligences (such as linguistic, logical-mathematical, spatial, musical, bodily-kinesthetic, naturalistic, interpersonal, and intrapersonal) and begin to see themselves as capable, competent mathematicians (Davis, Christodoulou, Seider, & Gardner, 2011). As I predict will be the case with your students, addressing underlying mathematical concepts led to students self-reporting how topics were starting to make sense, and they were "figuring it out" in their other mathematics classes. Similarly, those other mathematics teachers reported improvements in students' engagement, completion, and understanding of grade-level content as well. This was all the justification we needed to continue an approach that *listens* to the symptoms yet *addresses* the *causes*, rather than only the symptoms.

To Sum It Up

Integrate innovative approaches to learning that target the root of developing mathematical understanding.

When we reflect on our most discouraging personal or professional seasons, they likely share a common thread: *I am stuck, and I cannot figure out how to get unstuck.* The same is true for our students. Let's *consider* the factors outside our control, but *invest* our time in adjusting what is within our control. It is wise to speak up, believe in, and advocate for yourself as you build the CTE necessary to impact student achievement.

"Sure, Jenn," a colleague smirked after I offered my perspective about creative solutions to meeting student needs. "Just wait until you've taught a year. You'll change your tune." I felt a momentary pang of grief for what appeared to be a teacher who no longer believed in her power to make a difference. I don't know with certainty if this teacher ever felt enthusiastic about her ability to impact students or if she always felt the factors outside her control were insurmountable. Whichever the case, I heard her. I had felt it too, as I am sure you have. It's OK, normal even, to encounter this sense that we have hit a wall we cannot overcome.

However, with as much emphasis as I can possibly convey, *do not stop there.* Do not stop at the end of your *if-onlys.* Radical acceptance motivates us to press through to *because.* Embrace the challenges and intense realities with the belief that your efforts to find a way, to figure it out, and to adjust as needed, *do indeed make a difference.* You may not directly experience the fruits of your efforts for a while, if ever. Trust that you are planting seeds of engagement and belonging from which strong trees will grow. You are teaching students it is worth it to persevere, to focus on what is within their control. You are teaching students there is no shame in not knowing, needing help, or having a complex set of life circumstances. You are encouraging and empowering students to stay engaged in their truths and build what they need and want from where they are. After all, we are in the business of belonging!

Chapter 7 Application Guide

Use the reproducible "Chapter 7 Application Guide" (page 130) to gather your thoughts and make connections between the themes and tools in chapter 7 and your teaching practice. Consider using this application guide collaboratively with colleagues to anchor your exploration, discussions, and application of your learning.

Chapter 7 Application Guide

Chapter Themes and Tools	Connect and Apply to Your Practice
Moving from *If only . . .* to *Because . . .*	Listen to conversations between and among teachers and staff for the *if-only* statements that prevent them from finding necessary solutions. In your own reflection or with colleagues (if you are comfortable), rephrase the *if-only* to a *because* and see how that changes your next steps.
Reconsidering homework	What are your (or your team's) current homework practices? What leeway do you have to change that practice? What are your first steps to engage and possibly change your approach to homework?

page 1 of 2

Investigating underlying mathematical concepts	What skills do your students currently struggle with? What are the conceptual foundations of those skills? How do students develop those concepts from as early as kindergarten through a certain grade level? What resources might you need to deepen your own understanding of mathematical content, so you are better equipped to recognize the common denominators at the root of your students' struggles with grade-level content?

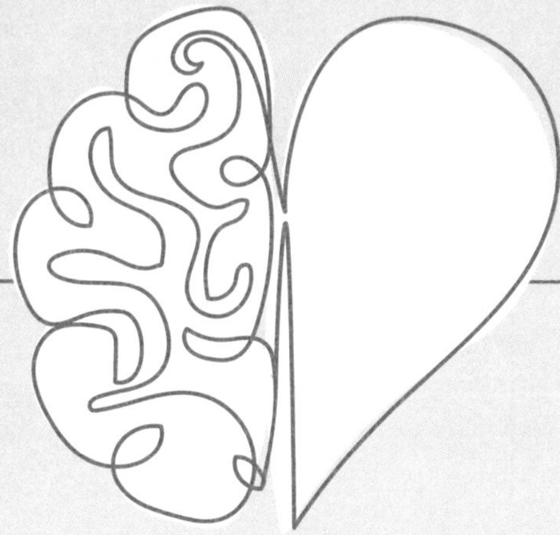

"What's the bravest thing

you've ever said?"

asked the boy.

"Help,"

said the horse.

—Charlie Mackesy

CHAPTER 8

I Have 250 Students Every Semester. I Can't Possibly . . .

Whether this is your first year of teaching or your tenth, whether you work in an urban or rural setting, whether you teach mathematics intervention or AP Calculus, you will experience some manner of fatigue along your journey. I know for many, fatigue is the most prevalent aspect of their current reality. It makes sense to experience deep fatigue given the unrelenting circumstances and events in our personal lives and, more broadly, in our profession. How common is it to hear the following exchange?

"Hey there! How are you?"

"I'm good! Busy. *Tired.* I have *so many* students. But good!"

I have certainly been there; surrendering to perpetual *busy* and perpetual *tired* as I strive to do everything I can to make a difference in this field of education. It's that "full steam ahead, pedal to the metal, all-out push every day" that masquerades as the ultimate evidence we're really giving it our all. In all honesty, however, this may not be congruent with a life we would describe as *good*, and certainly not one that is sustainable. We must aim to strike a better balance between the full exertion of "race day" and the training or rest days preceding and following the "main event." While we will likely not fully avoid every instance of getting worn out, we are wise to identify the factors that contribute to fatigue and plan ahead so we can position ourselves to be finishers in the "long-distance event" of teaching.

Knowing what you are up against is the only way to properly prepare. Without a comprehensive understanding of the terrain, how would you know what gear to bring or what to wear? How would you know how to train and prepare for your best performance on "race day"? So let's begin! In this final chapter, I ask you to join me for an honest analysis of the teaching "racecourse" I traversed and the lessons I learned along the way. I hope that by telling my story, you will recognize your own similar experiences, and we can together explore some solutions for dealing with those challenges.

What's Really Going on Here

When I landed my first teaching job, I remember feeling an "emotional cocktail" of relief and enthusiasm—*relief* to have a job at all, and *enthusiasm* to be among the inaugural group of mathematics-intervention teachers across the district. My enthusiasm waned slightly when I learned the district would mix sixth-, seventh-, and eighth-grade students across ten different class periods spanning the alternating two-day A/B block schedule. This meant I would be planning for, teaching, and assessing nearly 250 students each semester, with the possibility that student enrollment in particular class periods might shift every six weeks to accommodate other scheduling constraints. For example, if a student was in my third-period class for the first six weeks, that same student may be in my seventh-period mathematics intervention the next six weeks. The challenges these students would face building new relationships with new classmates (and the instructional obstacles this posed for me) were concerning. I was told this was the only option given the master schedule. Let's call this *factor one.*

My second piece of logistic news was I would be teaching in the portable units outside the main school building. I was not thrilled; this could remove me from most school happenings and distance me from fellow mathematics teachers. Teaching outside the main building gave me a sense of being in the *other* category, and I'm confident my students felt that way too. *Those students left the main building to go to mathematics intervention.* This was a singling out of sorts. This second factor compounded the already uphill battle to re-engage students in learning. From what I could tell, the school's decisions about which teachers taught in which classrooms were political, multidimensional, and not open for discussion.

When I look back at that setup through an equity lens, it's troublesome (NCTM, 2018). From a student perspective, the requirement they leave the building for intervention instruction also meant mathematics or reading interventions took the place of an elective in their schedule. If a student needed reading *and* mathematics intervention, they lost two electives. I submit to you *factor two.*

When I accepted this teaching position, I lived about sixty miles from the school. For two years, my workday included a daily 120-mile round-trip commute. For a time, I took the train, leaving my house each day around 5:00 a.m., arriving at the train station, boarding the 5:30 a.m. northbound train, and arriving at my destination around 6:45 a.m. The mathematics coach at our school picked me up on her way. Each afternoon, we would leave school, and she would take me to the train station for my trek home. Rinse and repeat for the entire school year, and then another.

There were benefits to this arrangement. While on the train, I could read student journals, correct assignments or assessments, plan upcoming lessons, or sleep. When the train was no longer the most viable option, I found a vanpool of state workers also making this commute. Each day, we would meet, jet up the highway, drop everyone at their respective workplaces, work our days, then reverse route and head home. I often drove the vanpool because I was the last stop on the drop-off route. This meant less time for the preparatory work of teaching but was more reliable timing. This commute, my friends, was my *factor three.*

Once we have a clear picture of the terrain ahead of us and are able to identify or anticipate the aspects of our journey that will be tiring, we can embark on our preparation and training, making sure we are mentally and physically ready. As we did in chapter 5 (page 79), let's use the Portland relay race as a metaphor and use the same chapter structure to unpack the aspects of our preparation for and participation in the "long-distance event" of teaching.

PAUSE AND PONDER

What personal and professional factors wear you down?

Drivers and Walkers

In chapter 5 (page 79), I described students as the walkers and teachers as the drivers who play a particular role in supporting the walkers in achieving their targeted race-day outcomes. For this chapter, however, shift gears and think of me (and *yourself*) as the walker, the event participant. From this perspective, several "drivers" play integral roles helping me (as the walker) throughout my years of teaching. For the first two years of this teaching assignment, my supporters (drivers) included a school-based mathematics coach, a district-based new teacher mentor, and a principal who believed in my abilities. The principal helped clear as many obstacles standing between my students and their learning (such as student licenses to a particular software program, class sizes, and so on) as she could. I admit I was initially resistant to the support of our mathematics coach and the new teacher mentor. I espoused an "I don't need help" attitude I am sure they found delightful to encounter. I'll never forget a few turning points in my relationship with each person—moments where it started to sink in that my coach and mentor were not there to tell me how to do my job, but rather, to support me in any way *I* chose. This was, after all, my "event."

In the two weeks leading up to the beginning of the school year, teachers were attending planning meetings, setting up their classrooms, designing lessons, analyzing rosters, and so on. It was in one of these meetings when my mathematics coach learned I regularly commuted via train. Without hesitation, she offered to pick me up from the train station each morning and transport me back each afternoon. After considering my options, I accepted her offer. It was during our shared ride time I learned more about her heart for others, advocacy for students, and how she came to be the mathematics coach for our school. A particular interaction with her is emblazoned in my memory. One early morning, as we walked in the wintry darkness toward the main school building, she asked me how she could support me that day. I deflected, declining the offer, and letting her know I was fine without her help.

"You know—being a mathematics coach isn't a job where I look at teachers and say, *I gotcha*," she said as she frowned, pointing her index finger in an accusatory gesture. "It's more like, *I gotcha*," she explained as she softened her gaze and cupped her hands together like you would if you were gently catching something as it falls. I knew immediately what she meant. She felt acutely I had received her interjections into my lessons or planning as though she was accusing me of missing details. Which, of course, made me more defensive of my approach rather than able to hear her suggestions as opportunities to enhance the great work I was already doing. This exchange—and her direct explanation of her coaching intentions—changed everything. It was the perspective shift I needed to let down some of the walls I had constructed and really start hearing her feedback as supportive. It was a humbling moment for me as I realized there was something I could learn from the perspective of another; I could be a more effective "walker" if I willingly enlisted the support of a "driver."

The second turning-point conversation was with the new teacher mentor the district had assigned to me. I say "assigned to me" because I did not think I needed a mentor. The blanket policy, *All teachers new to the district have a mentor*, felt demeaning, as if my expertise or skills were not good enough. Again, I am not sure if it was pride or a desire for autonomy that motivated my reluctance to be mentored. It seems silly, in hindsight, not to take all the help you can get when you are teaching. Nevertheless, I resisted.

After my mentor observed one of my lessons midway through my first year, we sat down to debrief. "How do you think it went?" I asked her, internally bracing myself for critique.

"How do *you* think it went?" she asked in response. This may seem insignificant, but in her question, she acknowledged my expertise, valued my perspective, and conveyed the important *on-your-sideness* I didn't even know I was craving. I had been so focused on what she might say about my teaching, I was caught off guard at the invitation to be my own expert. Thinking back on it now, she was a skilled mathematics teacher in that moment, and when her "student" (me) asked, "Did I get it right?" she did what all great mathematics teachers do and said to her student, "You tell me." I still get goosebumps thinking about it because, in that moment, I recognized how much she and I had in common in our approaches to teaching. Neither of us was in the business of *telling*. Both of us were in the business of helping others figure it out for themselves. This trust-building interaction, as well as the one with my mathematics coach, signaled to me these two women would gladly be "drivers" for me, knowing firsthand the hard work of being the "walker."

PAUSE AND PONDER

Who are your "drivers"?

Are you resistant to the idea of having a "driver"? Why?

Water Stations

Not all exhaustion or fatigue is preventable, but water stations along the way provide necessary opportunities for refueling. When I launched into teaching, I wasn't sure where I would need "water stations" because I wasn't sure what aspects of teaching would be the most exhausting. Honestly, I am not sure I thought I *would* need refueling opportunities. I brazenly assumed I could handle it—just as I had believed I did not need a mentor or an instructional coach. I was wrong, which came to light in a predictably shocking turn of events. There *were* parts of teaching that wore me out from the beginning. Hauling crates of student work home each evening and correcting papers all weekend was not sustainable, and I knew I needed to adjust or I wouldn't make it to the "finish line." In chapter 5 (page 79), we explored some "water stations" that can help students. Here, I share a few "water stations" that can help teachers, as they helped me.

WATER STATION ONE: USE SEATING CHARTS TO GATHER DATA

Create an aerial view template of your classroom desks arrangement and make a few blank copies to use for formative assessments throughout a lesson. Gather critical information about student understanding on a single page for easy reference as you plan future lessons. Figure 8.1 shows a sample seating chart template you might use for data collection.

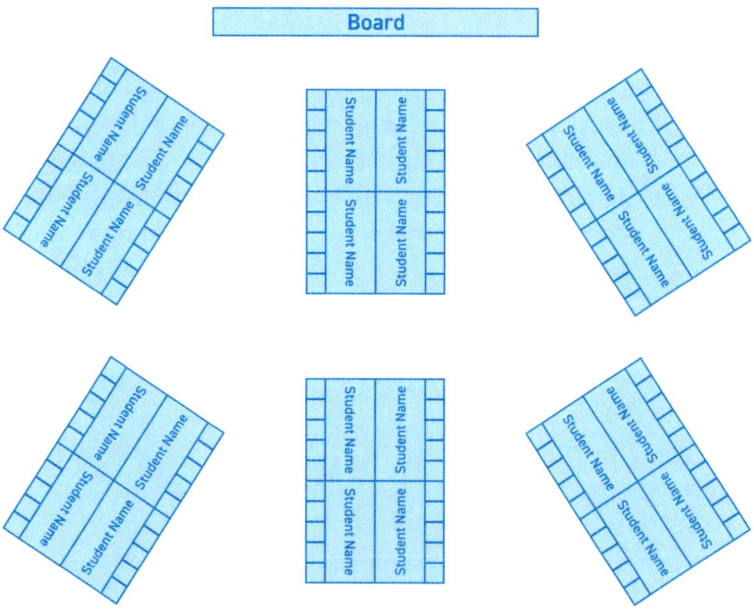

Figure 8.1: Sample seating chart template.

I modified the seating chart template (see figure 8.1) to include five small boxes adjacent to the rectangle representing each desk. This was space for me to record shorthand formative assessment information. Take a copy of the template and fill in student names where they are sitting for that week. Your seating chart should last about five class periods. My shorthand system included marking a box with an *A* when students were absent, leaving it blank if I didn't gather any information worth recording, or writing down a minus sign, check mark, or plus sign to indicate the student's demonstrated understanding of the key concepts or skills on that day.

- **Purpose:** This strategy serves two interrelated purposes.
  + It decreases the amount of paper corrections and analysis you need to do outside class.
  + It allows you to complete an analysis in real time and refer to it later, rather than relying on memory after a long day.
- **A consideration:** Circulating and gathering formative assessment data are most successful when students are engaged in collaborative working groups rather than direct instruction. Designing lessons to be student centered and student driven is a key component to freeing you up to listen and learn.

PAUSE AND PONDER

What aspects of your current practice occur during non-class time that you might accomplish during class instead?

WATER STATION TWO: INCLUDE STUDENTS IN THE CORRECTION PROCESS

Incorporate correcting student work into instructional time and engage students in the process of correcting *their own* as well as one another's work. Recall the test corrections example in chapter 6 (page 99). Additionally, you might have students collaborate, analyze, and correct common errors in small groups. Continue to invest in the social safety of your classes by anonymizing student work. Invite groups of students to consider and discuss the following questions.

- What understanding does this work example show?
- What support or questions can we ask this student to help clarify the student's thinking?
- How is this similar to or different from your own understanding?

Use think-alouds to teach students how to effectively engage in this process. "When feedback is delivered such that it is timely, specific, understandable, and actionable, students assimilate the language used by the teacher into their own self-talk" (Hattie et al., 2017, p. 203).

- **Purpose:** Including students in the correction process can reduce the amount of time you spend correcting work outside class. This process also results in prompt feedback for students on how they're doing and engages them in crucial self-monitoring and peer discussions. Structure lessons to include opportunities for students to study one another's work and self-correct as often as possible. This supports students in developing a positive mathematical identity, and helps them see themselves as capable problem solvers and their peers as thought partners in their sensemaking (Aguirre et al., 2013).
- **A consideration:** Students in mathematics intervention often believe mathematics is something only some people can understand. Structuring opportunities for self- and peer corrections is one way to dismantle this unproductive belief and show students mathematics is something they can indeed make sense of themselves. This does *not* mean I post correct answers and students check their work against mine. It means they compare their work with one another, notice where they disagree, and work together to better understand and reach resolution. The majority of students appreciate this genuine opportunity to share their knowledge and clarify their confusion with one another.

PAUSE AND PONDER

Think of an upcoming lesson. How might you engage students in correcting their own work or the work of one another?

WATER STATION THREE: SPEND LESS TIME REVIEWING WORK

Only review students' mathematics journals periodically, rather than daily or even weekly. Disperse the work of reviewing student journals across a period of time that works for your schedule. When paired with your regularly gathered formative assessment data, you can trust you won't miss any critical signs of student confusion. Consider a cadence, such as reviewing each student's journal about once every two weeks. Take this time to provide authentic written feedback on sticky notes you might add to students' journals. This periodic review gives you time to provide more comprehensive, student-specific feedback.

- **Purpose:** I found seeing *learning over time* helped me better understand a student's progress and learning trajectory than if I glanced at the student's work every day in a rush to "touch" every journal before the next class. This cadence of periodic reviews also allowed me to spend more time with each student's journal, leaving instructive feedback on sticky notes about the student's mathematical understanding as well as observations celebrating growth, persistence, concentration, and creativity I noticed in the student's work.

  A note about using sticky notes for leaving comments: students' journals belong to them. Encourage students to keep, move, or discard your sticky note comments as they see fit. Some students will prefer not to have a permanent interjection of their teacher's comments in the record of their learning, and students will appreciate having ownership of their journal.
- **A consideration:** I used journals (typically composition books) for ongoing work. When you're looking to gather more immediate snapshots of learning and want to see each student's thinking, consider using exit tickets on note cards instead of prompting students to respond in journals. An *exit ticket* is something you quickly collect as students leave class for the day. Balance the questions you ask on exit tickets to ensure you gather formative assessment information about students' learning of the mathematics as well as their attitudes about their learning experience that day.

PAUSE AND PONDER

How might you disperse the load of formal reviews of student learning across a period of time? What might be the benefits of this approach?

The ways in which our personal and professional worlds intertwine are important to consider. The water stations I described were causally related to my profession, but the impacts of integrating them into my daily routine impacted my personal life as well. Similarly, water stations I identified and used in my personal life impacted my professional well-being too. As you continue your journey, remember to consider the many aspects of your life, the varying stressors and challenges, and decide when and where a water station might be just the thing.

Encouragement

As I settled into my teaching groove, I met more teachers in the mathematics department and other domains. We met as school-organized content-area or grade-level collaborative teams. Whether meeting with the seventh-grade team of language arts, social studies, science, and mathematics teachers, or with our team of *mathies*, this structured opportunity for connecting with colleagues led to beneficial new relationships, a few of which I maintain today. These work-space relationships grew into friendships that thrived and encouraged me to persevere and recover from the most exhausting stretches of life. No matter the struggle, I found encouragement in the innovation of others. When the unrelenting onslaught of life challenges dimmed, I heard cheering from the sidelines, received high fives from teammates, and leaned on the energy of my fellow "walkers," who helped me keep pace and make it to the next resting point.

Since leaving the classroom, the prevalence of virtual and online learning communities has grown into something I never would have imagined. The beauty of an online or virtual community of supporters is found in the miles-apart-but-together-in-purpose synergy we may or may not have with our in-building colleagues. Because of the possibility of an ever-growing interconnectedness, we can find the people we need and experience a community of encouragers. We can also *be* encouragers in this online space, co-constructing the belonging we need through supporting one another on journeys that may look different on the surface, but are similar to our own.

Encouragement may look different for you than it did for me. What feels encouraging today may change tomorrow. How you contribute as an encourager in your different circles may also change over time. The importance of encouragement remains constant. Be sure to acquire and stay connected to your sources of encouragement. Teaching is a long-distance event, my friend. It helps to be cheered on!

Nourishment

Recall from chapter 5 (page 79) the idea of consuming food simply because it is good, not necessarily for its nutritional value. "Nourishment" for yourself during a day at school can look like any of the following.

- Taking part (or all) of your prep period to listen to your favorite music and walk the track instead of correcting work or planning lessons

- Having a cup of coffee with a colleague during an in-service day
- Putting a "taking a peaceful pause" sign on your door during a teacher workday to minimize social output and protect your quiet if needed
- Writing quick notes of gratitude for specific people in your building and putting them on their desks or in their mailboxes
- Building in a time-buffer after meetings to ensure you have time to eat, drink, and take a break

Nourishment (in the form of connection with other teachers) took intentional effort on my part because of the physical arrangement of our school campus. Initially, few in-building teachers ventured out to the portables for anything other than work-related conversations. Eventually though, as connections grew, these moments of connection grew increasingly frequent and were my "egg salad sandwiches"—or what I sought out after a brutally challenging "leg of the race."

PAUSE AND PONDER

What is your "nourishment"?

What do you look forward to to stave off the long-term fatigue that can make you want to quit the "race"?

Interval Training

The single most important structure in the evolution of my instructional practice was the regular implementation of learning station rotations. (I did not know of a mathematics workshop model at the time, but I think what I developed has many similarities.) First, I will explain how station rotations worked in Portable 137, then I will outline some of the key benefits of this approach and why I highly recommend it for mathematics instruction.

I divided the class into three groups. There were three main stations for them to rotate through over the course of three class periods. I saw my students every other day (due to the A/B block schedule). Class periods were fifty-five minutes; we spent the first and final five minutes of class as a whole group, and students did two twenty-minute stations (with about five minutes for transitions). Table 8.1 (page 144) shows how this looked in a given class period.

Table 8.1: Sample Station Rotation Schedule

	Rotation One	Rotation Two
Group A	Teacher time	Computer or individual
Group B	Small groups	Teacher time
Group C	Computer or individual	Small groups

For students to experience every station during every class period, the time they spent at each learning station seemed much too short to actually engage with mathematics, and too much time was spent on transitions. Instead, I structured the rotations so students only did two of the three stations within a given class period, hitting each station twice over the span of three class periods. Table 8.2 illustrates how I structured rotations over three class periods.

Table 8.2: Station Rotations Over Three Class Periods

	Class Period One		Class Period Two		Class Period Three	
	Rotation One	Rotation Two	Rotation One	Rotation Two	Rotation One	Rotation Two
Group A	Teacher time	Computer or individual	Small groups	Teacher time	Computer or individual	Small groups
Group B	Small groups	Teacher time	Computer or individual	Small groups	Teacher time	Computer or individual
Group C	Computer or individual	Small groups	Teacher time	Computer or individual	Small groups	Teacher time

In an effort to be transparent and generate student buy-in, share the plan with students so they can see when they will get to their favorite station, which will likely be different for each student. Notice how students' engagement through their less-favorite options changes when they know they have something to look forward to. Depending on the learning needs of the class, I might arrange my groups by grade level (remember, sixth-, seventh-, and eighth-grade students were mixed within each class period), how well they worked together, or some other configuration. "Some of the most effective teachers make good use of flexible groupings that shift partnerships . . . to foster collaborative and cooperative learning" (Hattie et al., 2017, p. 155). Keep your groups fluid, changing them often as you tackle new content and as students learn to work collaboratively with different peers.

It may take some time to set up this station rotation model, so teach students how to engage with the small-group task, how to log in to devices, and so on. If you have resources available (such as a mathematics coach or another staff support), enlist their support early on when students are first learning the structure. You want to teach students the necessary self-management skills and set them up for success, which may mean they need the support of another adult as they cement positive, productive behaviors. Remember how we should practice smaller intervals before we set out to walk a mile, or ten, or fifty? The same logic applies to teaching the norms about station rotations.

There are many benefits to the station rotation approach, including the following.

- It frees up students from the learning intensity required to overcome confusion in mathematics. Rotating offers built-in breaks, chances to move, chances to say "Hi" to a friend in another group, and so on.
- It provides students the much-needed opportunity to develop their skills as independent learners and problem solvers by disrupting patterns of "I'm stuck; I quit" as interdependent groups form and support one another.
- It provides multiple and varied opportunities for students to encounter the concepts and skills they're learning.
- It frees you up to focus your attention on the acute learning needs of a smaller group of students and provide the targeted, differentiated instructional support they need. In my classroom, we used a personalized adaptive software program during computer time. Periodically, I carved out five or ten minutes for students to play mathematics games or to write during their computer time.

How did station rotations work in concert with circulating to check for understanding? Great question! The adaptive software we utilized provided its own reports and feedback, so I limited my review of students working on computers to analyzing any glitches in the software or assisting with confusing questions. The teacher-time rotation provided me the opportunity to carefully examine student thinking and listen for what aspects of the content they solidly understood and which they had partially developed understanding. Sometimes, during teacher time, students each might take two minutes to tackle a problem on their own. During that time, I could grab my clipboard and seating chart, and check in with my students working in small groups.

It is important to note that I did not use the station rotation structure for every class period, every day. Some days, we engaged in whole-class instruction so I could introduce an activity or task that students then tackled in smaller groups. Some days, we took assessments or engaged in test corrections. The station rotation structure was, however, the solution I found for differentiating instruction while providing the individualized attention, support, and instruction some students needed.

Needs Assessment

In any official race, there are things you can control and things you cannot. We cannot control the weather, other racers, onlookers, or the course. We do get to pick our pace, our level of output, the music we listen to, and so on. We get to decide when to produce a burst of speed or opt for a more moderate pace, knowing the miles that stretch ahead. When we are deciding whether to ease up or push harder, we gather one key set of data points by paying attention to the signals our bodies send and responding accordingly.

How do you know you are on the path leading to unrecoverable, "race-ending" overwhelm? What are your personal signs or indications your exertion is outpacing your reserves? For me, evidence appears mentally, physically, and emotionally. Conscious awareness of my own struggle is often the last sign to register. Long before I *think* I'm feeling really tired, my body signals to me I have had enough. Some of these signals might include the following.

- Headache or migraine
- An oversized emotional response to a relatively inconsequential event
- Difficulty concentrating
- Brain stuck on repeat, making it hard to move forward, see alternative perspectives, be creative, or be present

We know the power of continuing to be students of our own experiences, increasing our self-awareness, and becoming more compassionate in response to noticing when we need to adjust or get help. Taking time to notice the *pain points* in our personal or professional lives is a key strategy to a more sustainable approach to teaching and living. This is a way we can cultivate our own engagement and belonging in the teaching profession.

PAUSE AND PONDER

What are your personal and professional pain points?

How does your mind or body tell you that you are "running on empty"?

What can you learn from those signals?

To Sum It Up

Provide students with the engaging opportunities they need using practices and structures that also support you on your journey.

The lessons I continue to learn about fatigue impact all aspects of my personal and professional life. No longer do I prize exhaustion as an indication of worthy contribution. Instead, I more consistently pause and look at upcoming "legs of the race," and think about how I can finish the "next leg" with enough in my "reserve tank" for the one after that. This is perhaps the most important lesson I have learned about the demands of teaching and being human in this world. The "race" is not over after this class period. It is not over after this week or this school year. There is, and will continue to be, one more "leg." The next "leg of my race" always starts sooner than I want it to—before I have had time to fully recover from the previous one.

With that in mind, I choose not to "leave it all on the course," knowing the high price of overexertion and the physical, mental, and emotional cost of "running on empty." Take care of yourself along your journey and encourage others to do the same by stopping at "water stations," finding "drivers" you trust, leaning on your teammates for encouragement, treating yourself to some good nourishment, and noticing and asking for what you need. You are a crucial member of our collective "relay" team, and I am cheering you on!

Chapter 8 Application Guide

Use the reproducible "Chapter 8 Application Guide" (page 148) to gather your thoughts and make connections between the themes and tools in chapter 8 and your teaching practice. Consider using this application guide collaboratively with colleagues to anchor your exploration, discussions, and application of your learning.

Chapter 8 Application Guide

Chapter Themes and Tools	Connect and Apply to Your Practice
Using water stations	Revisit your notes from this chapter's first Pause and Ponder section (page 135). Choose a few key words to jot down in your planner as reminders to check in with yourself and set a sustainable trajectory. The workload of a teacher is significant. How might you distribute your work over time? Be sure to share with and get ideas from colleagues, and ensure that you have refueling opportunities throughout the school year.
Providing nourishment	Reflect on what is especially nourishing for you, perhaps what you noted in the corresponding Pause and Ponder section (page 143). Mark in your calendar when and how you will regularly incorporate nourishment into the upcoming months.

page 1 of 2

Fostering encouragement	Who are your most ardent supporters and encouragers? Give yourself something to look forward to by setting up a chat or meeting with your encouragers.
Using interval training	If you have not yet incorporated station rotations regularly into your instruction, I cannot recommend it enough. Choose an approach and try it out! It may take some adjustments over time to make it work the best for your classes and content, but it is *worth it.*

page 2 of 2

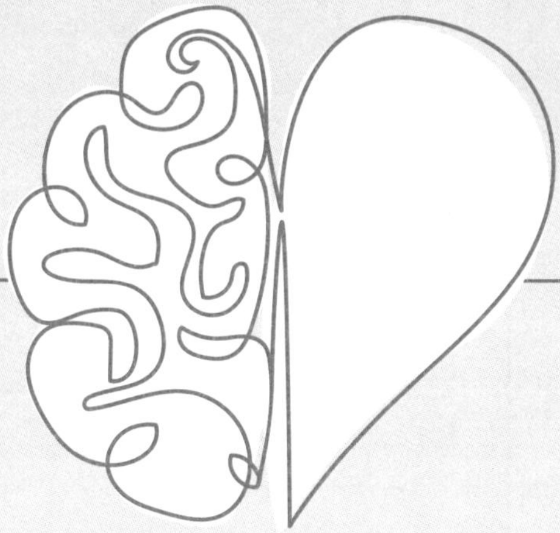

Become who you are.
Make what only you
can make.

—Friedrich Nietzsche

Epilogue

"What is *good teaching*?" asked Professor Rosiek, providing about two minutes for us to jot down our thoughts before discussing in small groups.

"What is *great teaching*?" he prompted us for a second round of thinking, writing, and then discussing.

"What is *beautiful teaching*?" he prompted a third time.

My heart skipped a beat. I knew I was in the right profession. I *knew*. Take a moment, right now. Close your eyes and answer the question, *What is beautiful teaching?* When you're ready, jot down your ideas. Keep your answer in a place you will remember and revisit when you encounter your own seasons of challenge, frustration, or despair. *Beautiful teaching* is what happens when we develop our conceptual understanding of the "denominator" (the whole student), and actively cultivate engagement and belonging. This is the outcome of our efforts, and it has a *beautiful* impact on learning. Trekking through the hard times, continuing to develop understanding of our common denominators, is what this journey is all about. You can make it.

Some days, we are good teachers. Some days, we are even great teachers. I hope this book equips you, in the most compelling way, for your pursuit of *beautiful teaching*—the kind of teaching that sets developing minds and souls on trajectories aimed beyond state test scores, achievement (as traditionally defined), and the constructs and constraints of what exists. We are in the business of making dreams come true, of believing in students when they may not believe in themselves. We are in the business of cultivating engagement and belonging. We must very loosely hold any single agenda, instructional routine, or lesson plan so that we can very tightly hold the *people* in our classes.

There will be cycles that challenge us to the very edge of our emotional and cognitive skill sets—guaranteed. Your students will feel it, and you will feel it. How you navigate those times and teach students to do the same speaks volumes about your investment in yourself and others. Take into consideration your wholeness and resist the fragmented narratives that reduce you to something less than your beautifully messy totality.

While there will be strife, there will *also* be persistence, creativity, innovation, and triumph. Through reimagining what might be and revisiting our most beautiful visions of teaching and learning, we can bring to life the high-quality learning our students deserve. Imagine a world where saying, "I'm not a math person," was obsolete. Imagine a world where we lean into the power of our collective efficacy and refuse to surrender to the story that we do not make a difference. Imagine, reader, just imagine!

Where you may have previously sought refuge in numbers and data about achievement, where you may have previously shied away from the big emotional experiences of middle and high schoolers, I encourage you to lean in. We cannot fix it all, but we can be present and acknowledge it all. It all belongs within our own experiences and in our classes. We have immeasurable power and influence to convey to students their challenges and setbacks are not final. Thank you for your ongoing commitment to yourself and students. You make an incredible difference. Together, we will keep innovating, encouraging, and maintaining a steadfast gaze on the beauty of this work as we cultivate our common denominators.

APPENDIX A

Research Influences

Authors, researchers, and speakers influenced my personal reflections and continue to influence my journey to understand how to foster mathematical belonging. I offer brief summaries of their work here and encourage you to seek out those that pique your interest. Additional works by these experts are included in the References and Resources section (page 165).

Social-Emotional Learning (SEL)

The Collaborative for Academic, Social, and Emotional Learning offers free ongoing webinars and many other resources for supporting social-emotional health and academic success (see https://casel.org). Cofounder and chair of CASEL, Timothy Shriver, writes, "emotion drives attention and attention drives learning" (CASEL, n.d.a). This encapsulates everything I found to be true when I taught, and what I continue to experience in my current role supporting teachers' professional development.

In addition to CASEL, be sure to check out Stephanie M. Jones's work through Harvard's Ecological Approaches to Social Emotional Learning (EASEL)

Laboratory (https://easel.gse.harvard.edu). Perhaps one of the most encouraging pieces of research Jones (2019) has shared is that SEL does not work best as a curriculum, but as skills and practices embedded into academic contexts. I find this encouraging because (as we all know) what teachers do *not* need is one more program. Jones's (2019) research confirms that high-quality teaching is a significant driver of student achievement.

Equity

Rochelle Gutiérrez's (2018) work on equity and rehumanizing mathematics continues to shape my understanding of classroom dynamics. She describes dimensions of equity, and reminds us the work to rehumanize mathematics is necessary because much of what happens in schools is, undeniably, dehumanizing (Gutiérrez, 2018). She calls attention to what we know to be true—that there is no such thing as a mathematics gene (Gutiérrez, 2018). Yet, whether others perceive you to be skilled mathematically has concrete, lasting, sometimes traumatic impacts on your life (Gutiérrez, 2018). Listening to or reading Gutiérrez's (2008, 2009, 2013, 2016, 2018) work reinforces my commitment to pursue beautiful teaching and break free from the educational constructs that marginalize and oppress.

Discretionary Spaces

Deborah Loewenberg Ball describes the *discretionary spaces* of teaching and learning when she writes about the critical importance of reflecting on and refining our practice so our actions in those discretionary spaces honor students (as cited in the American Educational Research Association, 2018). In her presidential address, Ball says:

> So discretion also has things that are very positive. It's through that discretion that teachers can get to know their communities, their children, their families. They can connect their teaching that they're doing to the environments in which they're working and so many people in this room do important work that shows what that can look like when it's done well. But it wouldn't be responsible for us not to recognize that discretion is also the source of a great deal of reproduction of racism because although teachers can connect school to the outside world it also allows racism and sexism that permeate our society and other forms of oppression—ways of thinking about gender ways of thinking about identity about ableism—all to come in through the discretionary acts that teachers are able to make and must make. (as cited in the American Educational Research Association, 2018)

The idea that *discretion* is both the gift and the challenge of teaching helps us with perspective taking. Discretion helps us understand we all have unconscious bias when it comes to gender, race, ableism, and so on, and we must be mindful—even vigilant—as we make decisions in those discretionary spaces. The aim is to disrupt, rather than repeat, systems of oppression. And we do that work often one moment at a time.

The Whole-Brain Child

Coauthors of *The Whole-Brain Child*, Daniel J. Siegel and Tina Payne Bryson (2011) write about how to engage with children in a way that teaches them relationships are about connection, communication, and compromise. Siegel and Bryson (2011) write about the associative machine in our brains we call "memory" (p. 67), and posit that "Neurons that fire together wire together" (p. 69). This means, for example, when students have a rough experience over something like the lack of a pencil, neurons related to that experience (including neurons about the teacher, peers, and class) wire together. Our memories then associate those feelings with the teacher, peers, and class.

Similarly, when teachers have a rough interaction with a student, neurons that fired together (who the student is, how the situation made us feel) also wire together. Siegel and Bryson (2011) explain how our past memories set up an expectation for future events. As you can imagine, this has significant implications for those of us teaching students who have had years of negative experiences in school. This, combined with their explanation of implicit and explicit memories, paints a compelling picture of how our students need our help, not our criticism, to regain a brain balance conducive to learning and thriving (Siegel & Bryson, 2011). The coauthors write, "We learn early in life to use our connections with reliable others to soothe our internal distress" (Siegel & Bryson, 2011, p. 126). The way I see it, we strive to be the "reliable others" for our students (Siegel & Bryson, 2011, p. 126). This book also offers clear explanations for different areas of the brain, such as the impact of the amygdala, and how we can get our prefrontal cortex back online.

Vulnerability and Resilience

Professor, author, and researcher Brené Brown's (2012) book *Daring Greatly* served as my formal introduction to work about vulnerability and shame resilience. Brown (2012) begins with this quote from Theodore Roosevelt:

> It is not the critic who counts; not the man who points out how the strong man stumbles or where the doer of deeds could have done them better. The credit belongs to the man who is actually in the arena, whose face is marred by dust and sweat and blood; who strives valiantly; . . . who at the best knows in the end the triumph of high achievement, and who at the worst, if he fails, at least fails while daring greatly. (p. xviii)

Daring Greatly speaks to my experiences of trying, failing, and trying again (Brown, 2012). A book about courage, Brown's (2012) writing gives us permission to explore our own pockets of vulnerability and reflect on the ways we can make space for others to explore their own vulnerabilities.

Her books *Braving the Wilderness* and *Dare to Lead* both offer strategies for wholehearted living (Brown, 2017, 2018). In *Dare to Lead,* Brown (2018) writes:

> In our original research on shame, 85 percent of the people we interviewed could recall a school incident from their childhood that was so shaming, it changed how they thought of themselves as learners. What makes this worse is that approximately half of those recollections were what I refer to as creativity scars. The research participants could point to a specific incident in which they were told or shown that they weren't good writers, artists, musicians, dancers, or something else creative. The shame tool used in these situations was almost always comparison.
>
> On the flip side of that finding, the same data showed that more than 90 percent of the people we interviewed could name a teacher, coach, school administrator, or faculty member who reinforced their self-worth and helped them believe in themselves and their ability. (p. 132)

Grading

Former NCTM President Matthew R. Larson's (2018) ongoing work sheds light on the detrimental impact of teacher and student tracking. In particular, his 2018 NCTM presentation about high school mathematics and the lasting impact of getting an *F* in mathematics impacted my understanding of the systems we are working to overhaul as we disrupt exclusionary patterns in schools.

In the book *Catalyzing Change in High School Mathematics*, the NCTM (2018) writes about the impact of student tracking and decisions to place students in advanced, regular, or low-level mathematics courses. The grades students receive in middle school impact their placement in high school mathematics programs. This conversation is crucial to understanding the importance of eliminating conventional grading practices.

The Trouble With the Achievement Gap

In her article "Embracing the Inherent Tensions in Teaching Mathematics From an Equity Stance," Gutiérrez (2009) writes:

> The excessive focus that U.S. researchers place on the gap between the mathematics achievement of White, middle-class students and that of African American, Latina/Latino, American Indian, and working class students, as well as English language learners, and the need to close the gap (termed "gap gazing") sheds light on issues of access and achievement from a dominant perspective with little concern for issues of identity and power or broadened notions of learning from a critical perspective. (p. 9)

This article explains how our focus on the achievement gap influences our teaching strategies (Gutiérrez, 2009). It also calls for teacher education programs to shift the conversation away from the achievement gap "to one that more easily exposes complex issues of identity and power and better reflects the interpretive activity involved in learning to teach" (Gutiérrez, 2009, p. 10). In conclusion, Gutiérrez (2009) writes, "We need to see our work as teacher educators as also supporting teachers to develop a counter-narrative to limited discourses like the achievement gap that can alienate teachers from their wisdom and disconnect research from practice" (p. 15).

APPENDIX B

Sources for Mathematics Lessons and Professional Learning

In this appendix, I compiled a list of books that offer high-quality lessons and tasks, and can aid in professional development.

Number Sense

- *It Makes Sense!* series by Melissa Conklin (2022), Ann Carlyle (2022), and Melissa Conklin and Stephanie Sheffield (2022)
- *Figuring Out Fluency in Mathematics Teaching and Learning, Grades K–8: Moving Beyond Basic Facts and Memorization* by Jennifer Bay-Williams and John SanGiovanni (2021)
- *Number Talks: Whole Number Computation* by Sherry Parrish (2022)
- *Number Talks: Fractions, Decimals, and Percentages* by Sherry Parrish and Ann Dominick (2022)

High-Quality Tasks

- *Lessons for Algebraic Thinking* series
  + *Grades K–2* by Leyani von Rotz and Marilyn Burns (2002)
  + *Grades 3–5* by Maryann Wickett, Katharine Kharas, and Marilyn Burns (2002)
  + *Grades 6–8* by Ann Lawrence and Charlie Hennessy (2002)
- *Active Algebra: Strategies and Lessons for Successfully Teaching Linear Relationships, Grades 7–10* by Dan Brutlag (2009)
- *Good Questions for Math Teaching: Why Ask Them and What to Ask* series
  + *Grades K–5* by Peter Sullivan and Pat Lilburn (2022)
  + *Grades 5–8* by Lainie Schuster and Nancy Anderson (2022)
  + *High School* by Nancy Anderson, Leslie Dietiker, and Gregg Reilly (2022)
- *Beyond Pizzas and Pies: Ten Essential Strategies for Supporting Fraction Sense, Grades 3–5* by Julie McNamara and Meghan M. Shaughnessy (2022)
- *Beyond Invert and Multiply: Making Sense of Fraction Computation, Grades 3–6* by Julie McNamara (2022)

Mathematical Discourse and Multilingual Learners

- *Classroom Discussions: Using Math Talk to Help Students Learn, Grades K–6* by Suzanne Chapin, Catherine O'Connor, and Nancy Canavan Anderson (2009)
- *Supporting English Language Learners in Math Class, Grades 6–8* by Kathy Melanese, Luz Chung, and Cheryl Forbes (2011)

Adolescent Development

- *Adolescents at School: Perspectives on Youth, Identity, and Education* edited by Michael Sadowski (2020)
- *The Deepest Well: Healing the Long-Term Effects of Childhood Adversity* by Nadine Burke Harris (2018)

Teaching

- *Culturally Responsive Teaching and the Brain* by Zaretta Hammond (2015)
- *Faster Isn't Smarter* by Cathy Seeley (2015)
- *For White Folks Who Teach in the Hood . . . and the Rest of Y'all Too: Reality Pedagogy and Urban Education* by Christopher Emdin (2016)
- *All Students Must Thrive* by Tyrone C. Howard (2019)

References and Resources

Aguirre, J., Mayfield-Ingram, K., & Martin, D. B. (2013). *The impact of identity in K–8 mathematics learning and teaching: Rethinking equity-based practices.* Reston, VA: National Council of Teachers of Mathematics.

American Educational Research Association. (2018, May 4). *AERA 2018 presidential address: Deborah Loewenberg Ball* [Video file]. Accessed at https://youtube.com/watch?v=JGzQ7O_SIYY&t=3s on August 30, 2022.

Anderson, N., Dietiker, L., & Reilly, G. (2022). *Good questions for math teaching: Why ask them and what to ask, high school.* Portsmouth, NH: Heinemann.

Arthur, Y. D., Owusu, E. K., Asiedu-Addo, S., & Arhin, A. K. (2018). Connecting mathematics to real life problems: A teaching quality that improves students' mathematics interest. *Journal of Research and Method in Education, 8*(4), 65–71.

The Aspen Institute National Commission on Social, Emotional, and Academic Development. (2019, January 15). *From a nation at risk to a nation at hope.* Accessed at https://files.eric.ed.gov/fulltext/ED606337.pdf on August 29, 2022.

Barbieri, C. A. & Miller-Cotto, D. (2021). The importance of adolescents' sense of belonging to mathematics for algebra learning. *Learning and Individual Differences,* (87). https://doi.org/10.1016/j.lindif.2021.101993

Bay-Williams, J. M., & SanGiovanni, J. J. (2021). *Figuring out fluency in mathematics teaching and learning, grades K–8: Moving beyond basic facts and memorization.* Thousand Oaks, CA: Corwin Press.

Blakemore, E. (2018, March 27). *How dolls helped win Brown v. Board of Education.* Accessed at https://history.com/news/brown-v-board-of-education-doll-experiment on August 29, 2022.

Boaler, J. (2008). Promoting "relational equity" and high mathematics achievement through an innovative mixed-ability approach. *British Educational Research Journal, 34*(2), 167–194.

Boaler, J. (2022). *Mathematical mindsets: Unleashing students' potential through creative mathematics, inspiring messages and innovative teaching* (2nd ed.). Hoboken, NJ: Jossey-Bass.

Boaler, J., & Staples, M. (2008). Creating mathematical futures through an equitable teaching approach: The case of Railside School. *Teachers College Record, 110*(3), 608–645.

Boaler, J., Wiliam, D., & Zevenbergen, R. (2000). *The construction of identity in secondary mathematics education.* Accessed at https://researchgate.net/publication/29452435_The_Construction_of_Identity_in_Secondary_Mathematics_Education on November 13, 2023.

Brown, B. (2012). *Daring greatly: How the courage to be vulnerable transforms the way we live, love, parent, and lead.* New York: Avery.

Brown, B. (2017). *Braving the wilderness: The quest for true belonging and the courage to stand alone.* New York: Random House.

Brown, B. (2018). *Dare to lead: Brave work. Tough conversations. Whole hearts.* New York: Random House.

Brown, B. (Host). (2022a, February 14). Brené with Dan Pink on the power of regret [Audio podcast episode]. In *Dare to Lead.* Accessed at https://brenebrown.com/podcast/the-power-of-regret on August 3, 2023.

Brown, M. (2022b, March 31). *What radical acceptance really means* [Blog post]. Accessed at https://psychcentral.com/blog/what-it-really-means-to-practice-radical-acceptance on October 28, 2023.

Brutlag, D. (2009). *Active algebra: Strategies and lessons for successfully teaching linear relationships, grades 7–10.* Sausalito, CA: Math Solutions.

Carlyle, A. (2022). *It makes sense! Using number paths and number lines to build number sense, grades K–2.* Portsmouth, NH: Heinemann.

Chapin, S. H., O'Connor, C., & Anderson, N. C. (2009). *Classroom discussions: Using math talk to help students learn, grades K–6* (2nd ed.). Sausalito, CA: Math Solutions.

Cohen, E. G., Lotan, R. A., Scarloss, B. A., & Arellano, A. R. (1999). Complex instruction: Equity in cooperative learning classrooms. *Theory Into Practice, 38*(2), 80–86. https://doi.org/10.1080/00405849909543836

Collaborative for Academic, Social, and Emotional Learning. (n.d.a). *Advancing social and emotional learning.* Accessed at https://casel.org/ on December 4, 2023.

Collaborative for Academic, Social, and Emotional Learning. (n.d.b). *What does the research say? Hundreds of independent studies confirm: SEL benefits students.* Accessed at https://casel.org/fundamentals-of-sel/what-does-the-research-say on August 28, 2022.

Conklin, M. (2022). *It makes sense! Using ten-frames to build number sense, grades K–2.* Portsmouth, NH: Heinemann.

Conklin, M., & Sheffield, S. (2022). *It makes sense! Using the hundreds chart to build number sense, grades K–2.* Portsmouth, NH: Heinemann.

Cooke, A. N., & Halberstadt, A. G. (2021). Adultification, anger bias, and adults' different perceptions of Black and White children. *Cognition and Emotion, 35*(7), 1416–1422. https://doi.org/10.1080/02699931.2021.1950127

Cuellar, A. E., & Markowitz, S. (2015). School suspension and the school-to-prison pipeline. *International Review of Law and Economics, 43,* 98–106. https://doi.org/10.1016/j.irle.2015.06.001

Darling-Hammond, L., Flook, L., Cook-Harvey, C., Barron, B., & Osher, D. (2020). Implications for educational practice of the science of learning and development. *Applied Developmental Science, 24*(2), 97–140. https://doi.org/10.1080/10888691.2018.1537791

Darling-Hammond, S. (2023). *Fostering belonging, transforming schools: The impact of restorative practices.* Palo Alto, CA: Learning Policy Institute. Accessed at https://learningpolicyinstitute.org/media/4046/download?inline&file=Restorative_Practices_REPORT.pdf on February 1, 2024.

Davis, K., Christodoulou, J., Seider, S., & Gardner, H. (2011). The theory of multiple intelligences. In R. J. Sternberg & S. B. Kaufman (Eds.), *Cambridge handbook of intelligence* (pp. 485–503). Cambridge, England: Cambridge University Press. Accessed at https://papers.ssrn.com/sol3/papers.cfm?abstract_id=2982593 on November 13, 2023.

Duckworth, A. (2016). *Grit: The power of passion and perseverance.* New York: Scribner.

DuFour, R., DuFour, R., Eaker, R., Many, T. W., & Mattos, M. (2016). *Learning by doing: A handbook for Professional Learning Communities at Work* (3rd ed.). Bloomington, IN: Solution Tree Press.

Dweck, C. S. (2008). *Mindset: The new psychology of success.* New York: Random House.

Emdin, C. (2016). *For White folks who teach in the hood . . . and the rest of y'all too: Reality pedagogy and urban education.* Boston: Beacon Press.

Fernandes, A., Crespo, S., & Civil, M. (Eds.). (2017). *Access and equity: Promoting high-quality mathematics in grades 6–8.* Reston, VA: National Council of Teachers of Mathematics.

Flores, A. (2007). Examining disparities in mathematics education: Achievement gap or opportunity gap? *The High School Journal, 91*(1), 29–42. https://doi.org/10.1353/hsj.2007.0022

Fosnot, C. T., & Dolk, M. (2001). *Young mathematicians at work: Constructing number sense, addition, and subtraction.* Portsmouth, NH: Heinemann.

Goff, P. A., Jackson, M. C., Di Leone, B. A. L., Culotta, C. M., & DiTomasso, N. A. (2014). The essence of innocence: Consequences of dehumanizing Black children. *Journal of Personality and Social Psychology, 106*(4), 526–545. https://doi.org/10.1037/a0035663

Gutiérrez, R. (2008). A "gap-gazing" fetish in mathematics education? Problematizing research on the achievement gap. *Journal for Research in Mathematics Education, 39*(4), 357–364.

Gutiérrez, R. (2009). Embracing the inherent tensions in teaching mathematics from an equity stance. *Democracy and Education, 18*(3), 9–16.

Gutiérrez, R. (2013). The sociopolitical turn in mathematics education. *Journal for Research in Mathematics Education, 44*(1), 37–68. https://doi.org/10.5951/jresematheduc.44.1.0037

Gutiérrez, R. (2016). Strategies for creative insubordination in mathematics teaching. *Teaching for Excellence and Equity in Mathematics, 7*(1), 52–60. Accessed at www.todos-math.org/assets/documents/TEEM/teem7_final1.pdf#page=52 on August 28, 2022.

Gutiérrez, R. (2018, March 8–10). *Rehumanizing mathematics: A vision for the future* [Conference presentation]. Latinx in Mathematical Sciences Conference, Los Angeles, CA.

Hammond, Z. (2015). *Culturally responsive teaching and the brain: Promoting authentic engagement and rigor among culturally and linguistically diverse students.* Thousand Oaks, CA: Corwin.

Harris, N. B. (2018). *The deepest well: Healing the long-term effects of childhood adversity.* Boston: Houghton Mifflin Harcourt.

Hathaway, B. (2016, September 27). Implicit bias may explain high preschool expulsion rates for Black children. *Yale News.* Accessed at https://news.yale.edu/2016/09/27/implicit-bias-may-explain-high-preschool-expulsion-rates-black-children on August 29, 2022.

Hattie, J., Fisher, D., & Frey, N. (2017). *Visible learning for mathematics, grades K–12: What works best to optimize student learning.* Thousand Oaks, CA: Corwin Mathematics.

Howard, T. C. (2019). *All students must thrive: Transforming schools to combat toxic stressors and cultivate critical wellness.* Rexford, NY: International Center for Leadership in Education.

Hydon, S., Wong, M., Langley, A. K., Stein, B. D., & Kataoka, S. H. (2015). Preventing secondary traumatic stress in educators. *Child and Adolescent Psychiatric Clinics of North America, 24*(2), 319–333. https://doi.org/10.1016/j.chc.2014.11.003

Institute for Pure and Applied Mathematics (IPAM). (2018, March 16). *Rochelle Gutiérrez: Rehumanizing mathematics—A vision for the future* [Video file]. Accessed at https://youtube.com/watch?v=D266LYIigS0 on August 30, 2022.

Jones, S. M. (2019, June 25). *Three big ideas for effective SEL initiatives based on the science and practice of SEL* [Conference presentation]. Model Schools Conference, Washington, DC. Accessed at https://handouts19.modelschoolsconference.com/files/upload/Jones_MSC_June25_v1.pdf on February 1, 2024.

Joyner, J. M., & Bright, G. W. (2016). *INFORMative assessment: Formative assessment to improve mathematics achievement, middle and high school*. Sausalito, CA: Math Solutions.

Kaplinsky, R. (n.d.). *Lessons*. Accessed at https://robertkaplinsky.com/lessons on August 30, 2022.

Kuo, M. (2016, July 22). Low math confidence discourages female students from pursuing STEM disciplines. *Science*. https://doi.org/10.1126/science.caredit.a1600110

Ladson-Billings, G. (2006, April 26). From the achievement gap to the education debt: Understanding achievement in U.S. schools. *Educational Researcher, 35*(7), 3–12.

Lawrence, A., & Hennessy, C. (2002). *Lessons for algebraic thinking, grades 6–8*. Sausalito, CA: Math Solutions.

Ladson-Billings, G. (2014). Culturally relevant pedagogy 2.0: A.k.a. the remix. *Harvard Educational Review, 84*(1), 74–84. https://doi.org/10.17763/haer.84.1.p2rj131485484751

Larson, M. R. (2018). *Catalyzing change in high school mathematics: Initiating critical conversations* [Conference presentation]. NCTM Annual Meeting and Exposition, Washington, DC.

Lee, S. J. (2009). *Unraveling the "model minority" stereotype: Listening to Asian American youth* (2nd ed.). New York: Teachers College Press.

Lewis, D. (n.d.a). *Confederated tribes of Grand Ronde*. Accessed at https://oregonencyclopedia.org/articles/confederated_tribes_of_grand_ronde/#.Yw-u_HbMLD5 on August 31, 2022.

Lewis, D. (n.d.b). *Willamette Valley treaties*. Accessed at https://oregonencyclopedia.org/articles/willamette_valley_treaties/#.Yw-k_HbMLD4 on August 31, 2022.

Lian, B., Kristiawan, M., Primasari, D. A. G., & Prasetyo, M. A. M. (2020). Teachers' model in building students' character. *Journal of Critical Reviews, 7*(14), 927–932.

Mackesy, C. (2019). *The boy, the mole, the fox and the horse*. New York: Ebury Press.

Maier, E. (n.d.). *Num•ber sense/numb•er sense*. Accessed at https://mathlearningcenter.org/curriculum/free/additional/gene/number_sense_number_sense on August 29, 2022.

Marzano, R. J. (2019). *Understanding rigor in the classroom*. York, PA: Learning Sciences International.

Matthews, J. S. (2018). When am I ever going to use this in the real world? Cognitive flexibility and urban adolescents' negotiation of the value of mathematics. *Journal of Educational Psychology, 110*(5), 726–746. https://doi.org/10.1037/edu0000242

McGrew, K. (2016). The dangers of pipeline thinking: How the school-to-prison pipeline metaphor squeezes out complexity. *Educational Theory, 66*(3), 341–367. https://doi.org/10.1111/edth.12173

McNamara, J. (2022). *Beyond invert and multiply: Making sense of fraction computation, grades 3–6*. Portsmouth, NH: Heinemann.

McNamara, J., & Shaughnessy, M. M. (2022). *Beyond pizzas and pies: Ten essential strategies for supporting fraction sense, grades 3–5* (2nd ed.). Portsmouth, NH: Heinemann.

Melanese, K., Chung, L., & Forbes, C. (2011). *Supporting English language learners in math class, grades 6–8*. Sausalito, CA: Math Solutions.

Morris, M. W. (2016). *Pushout: The criminalization of Black girls in schools*. New York: New Press.

National Center for Education Statistics. (2019, February). *Indicator 15: Retention, suspension, and expulsion*. Accessed at https://nces.ed.gov/programs/raceindicators/indicator_rda.asp on August 29, 2022.

National Council of Teachers of Mathematics. (2014). *Principles to actions: Ensuring mathematical success for all*. Reston, VA: Author.

National Council of Teachers of Mathematics. (2018). *Catalyzing change in high school mathematics: Initiating critical conversations*. Reston, VA: Author.

Obama, M. (2018). *Becoming*. New York: Crown.

Oregon Department of Education. (n.d.). *Discipline, restraint and seclusion collections.* Accessed at https://oregon.gov/ode/students-and-family/healthsafety/Pages/DisciplineRestraintSeclusionCollections.aspx on August 29, 2022.

Paris, D. (2012). Culturally sustaining pedagogy: A needed change in stance, terminology, and practice. *Educational Researcher, 41*(3), 93–97.

Park, J., & Brannon, E. M. (2013). Training the approximate number system improves math proficiency. *Psychological Science, 24*(10), 2013–2019. https://doi.org/10.1177/0956797613482944

Parrish, S. (2022). *Number talks: Whole number computation.* Portsmouth, NH: Heinemann.

Parrish, S., & Dominick, A. (2022). *Number talks: Fractions, decimals, and percentages.* Portsmouth, NH: Heinemann.

Petroni, G. (2018, March 22). *Black girls perceived, often treated as older and less innocent, study finds.* Accessed at https://news.medill.northwestern.edu/chicago/black-girls-perceived-often-treated-as-older-and-less-innocent-study-finds on August 29, 2022.

Sadowski, M. (Ed.). (2020). *Adolescents at school: Perspectives on youth, identity, and education* (3rd ed.). Cambridge, MA: Harvard Education Press.

Schuster, L., & Anderson, N. (2022). *Good questions for math teaching: Why ask them and what to ask, grades 5–8* (2nd ed.). Portsmouth, NH: Heinemann.

Seeley, C. L. (2015). *Faster isn't smarter: Messages about math, teaching, and learning in the 21st century* (2nd ed.). Sausalito, CA: Math Solutions.

Siegel, D. J., & Bryson, T. P. (2011). *The whole-brain child: Twelve revolutionary strategies to nurture your child's developing mind.* New York: Delacorte Press.

Strong, L., & Das, A. (Hosts). (2018, December 11). Rehumanizing mathematics with Dr. Rochelle Gutiérrez [Audio podcast episode]. In *Abolition Science.* Accessed at https://abolitionscience.org/home/2018/12/11/rehumanizing-math-and-science-with-dr-rochelle-gutierrez on August 30, 2022.

Sullivan, P., & Lilburn, P. (2022). *Good questions for math teaching: Why ask them and what to ask, grades K–5* (2nd ed.). Portsmouth, NH: Heinemann.

Sultan, A. (2015, December 2). The limitations of teaching "grit" in the classroom. *The Atlantic.* Accessed at https://theatlantic.com/education/archive/2015/12/when-grit-isnt-enough/418269 on August 28, 2022.

Thomas, C. A., & Berry, R. Q., III. (2019). A qualitative metasynthesis of culturally relevant pedagogy and culturally responsive teaching: Unpacking mathematics teaching practices. *Journal of Mathematics Education at Teachers College, 10*(1), 21–30. https://doi.org/10.7916/jmetc.v10i1.1668

Van de Walle, J. A., Bay-Williams, J. M., Lovin, L. H., & Karp, K. S. (2018). *Teaching student-centered mathematics: Developmentally appropriate instruction for grades 6–8* (3rd ed.). New York: Pearson.

Visible Learning. (n.d.). *Collective teacher efficacy (CTE) according to John Hattie.* Accessed at https://visible-learning.org/2018/03/collective-teacher-efficacy-hattie on August 30, 2022.

von Neumann, J., & Morgenstern, O. (2007). *Theory of games and economic behavior* (60th anniversary ed.). Princeton, NJ: Princeton University Press. (Original work published 1944)

von Rotz, L., & Burns, M. (2002). *Lessons for algebraic thinking, grades K–2.* Sausalito, CA: Math Solutions.

Wickett, M., Kharas, K., & Burns, M. (2002). *Lessons for algebraic thinking, grades 3–5.* Sausalito, CA: Math Solutions.

Yu, T. (2006). Challenging the politics of the "model minority" stereotype: A case for educational equality. *Equity and Excellence in Education, 39*(4), 325–333. https://doi.org/10.108/10665680600932333

Index

Global PD
teams
Collaborative Learning
for School Improvement